INTRODUCTION TO TIME
SERIES ANALYSIS

Quantitative Applications in the Social Sciences

A SAGE PUBLICATIONS SERIES

INTRODUCTION TO TIME SERIES ANALYSIS

Mark Pickup

Simon Fraser University

Los Angeles | London | New Delhi
Singapore | Washington DC

Los Angeles | London | New Delhi
Singapore | Washington DC

FOR INFORMATION:

SAGE Publications, Inc.
2455 Teller Road
Thousand Oaks, California 91320
E-mail: order@sagepub.com

SAGE Publications Ltd.
1 Oliver's Yard
55 City Road
London, EC1Y 1SP
United Kingdom

SAGE Publications India Pvt. Ltd.
B 1/I 1 Mohan Cooperative Industrial Area
Mathura Road, New Delhi 110 044
India

SAGE Publications Asia-Pacific Pte. Ltd.
3 Church Street
#10-04 Samsung Hub
Singapore 049483

Copyright © 2015 by SAGE Publications, Inc.

Printed in the United States of America

Library of Congress Cataloging-in-Publication Data

Pickup, Mark (Associate professor)

Introduction to time series analysis / Mark Pickup, Simon Fraser University, Canada.

pages cm

Includes bibliographical references and index.

ISBN 978-1-4522-8201-5 (pbk. : alk. paper)

1. Time-series analysis. 2. Social sciences—Statistical methods. I. Title.

HA30.3.P53 2015

519.5'5—dc23   2014031504

This book is printed on acid-free paper.

Acquisitions Editor: Helen Salmon
Editorial Assistant: Anna Villarruel
Production Editor: Olivia Weber-Stenis
Copy Editor: Quads Prepress (P) Ltd
Typesetter: Hurix Systems Pvt. Ltd.
Proofreader: Jennifer Grubba
Indexer: Jeanne Busemeyer
Cover Designer: Candice Harman
Marketing Manager: Nicole Elliott

SFI Certified Sourcing
www.sfiprogram.org
SFI-00453

15 16 17 18 10 9 8 7 6 5 4 3 2

CONTENTS

SERIES EDITOR'S INTRODUCTION

Most commonly used statistical methods assume independently sampled observations. Dependencies among observations can arise for a number of reasons. For example, clustering in survey sampling can introduce non-negligible dependencies among the observations in a sample. Dependencies also arise naturally in data that are collected over time. In what are typically termed *longitudinal* data, a number of individuals is followed over time, and it is unlikely that repeated observations on the same individual are statistically independent. Common methods for analyzing longitudinal data include mixed-effects models, described, for example, in the QASS monographs by Luke (2004) and Preacher et al. (2008).

Time-series data differ from longitudinal data in that in the former a single unit of observation — typically an aggregated unit, such as a country — is followed over time, usually at regular intervals. In *univariate* time-series analysis, the subject of the current monograph by Mark Pickup, one time series is treated as a response or dependent variable. The object of the analysis is typically to relate this time series to others, which are treated as explanatory or independent variables, as well as to its own past values. The dependent character of the data complicates the analysis in comparison to regression methods for independent observations, as does the possible presence of previous values of the response variable on the right-hand side of the regression model, but these characteristics also offer opportunities for teasing out relationships that evolve dynamically over time.

Prof. Pickup broadly presents univariate time-series analysis, developing the topic from first principles. He introduces basic ideas, such as *autocorrelation*, *stationarity*, *exogeneity*, *trend*, and *periodicity*, prior to presenting the most widely used and useful univariate time- series models. This extensive treatment begins with time-series models in which the errors are stationary and serially uncorrelated, models that are estimable by ordinary least-squares regression. Prof. Pickup proceeds to introduce gradually and systematically a variety of complications, treating (among others) *autoregressive distributed-lag (ADL)* models; *lagged dependent-variable (LDV)* models; *autoregressive conditional heteroscedasticity (ARCH)* and *generalized autoregressive conditional heteroscedasticity (GARCH)* models; *autoregressive moving average (ARMA)* and *autoregressive integrated moving average (ARIMA)* models; and *error correction models (ECMs)*.

What distinguishes Prof. Pickup's from other presentations of time-series analysis are its lucidity and accessibility, along with his focus on the causal interpretation of theory- motivated time-series models — a focus that most social scientists will find more congenial than the emphasis on data-driven predictive models that motivates much of the time-series literature. The statistical content of the monograph is illustrated with an impressive variety of effective, judiciously selected, and carefully developed examples, and is complemented by a web site containing data and computer code for the various examples.

Time-series analysis has a reputation of difficulty, and Prof. Pickup doesn't shy away from addressing subtle and complex aspects of the topic; indeed, he has a knack for rendering inherently difficult material intuitively and clearly. As a consequence of its quality, coverage, orientation, and accessibility, I expect that this monograph will become a standard reference on time-series analysis for social scientists.

Prof. Pickup's monograph joins several others in the Sage QASS series that deal with various aspects of time-series analysis: It effectively updates the presentation of time-series regression methods by Ostrom (1990), and complements the presentation of univariate tests for time-series models by Cromwell et al. (1994b). *Multiple time series* — where several series are treated as possibly co-dependent response variables — are the subject of monographs by Brandt and Williams (2007), describing *vector autoregression (VAR)* models, and by Cromwell et al. (1994a), describing tests for multivariate time series. Prof. Pickup also deals in passing with *interrupted time series* analysis, which focuses on the effect on a series of an intervention or other event, and which is the subject of an early QASS monograph by McDowall et al. (1980).

—John Fox
Series Editor

References

Brandt, P. T. and Williams, J. T. (2007). *Multiple time series models*. Thousand Oaks CA.

Cromwell, J. B., Hannan, M. J., Labys, W. C., and Terraza, M. (1994a). *Multivariate tests for time series models*. Thousand Oaks CA.

Cromwell, J. B., Labys, W. C., and Terraza, M. (1994b). *Univariate tests for time series models*. Thousand Oaks CA.

Luke, D. A. (2004). *Multilevel modeling*. Thousand Oaks CA.

McDowall, D., McCleary, R., Meidinger, E. E., and Hay, Jr., R. A. (1980). *Interrupted time series analysis*. Thousand Oaks CA.

Ostrom, Jr., C. W. (1990). *Time series analysis: Regression techniques*. Thousand Oaks CA, second edition.

Preacher, K. J., Wichman, A. L., MacCallum, R. C., and Briggs, N. E. (2008). *Latent growth curve modeling*. Thousand Oaks CA.

PREFACE

This introductory text on time series analysis is a product of the time series analysis courses that I have taught both at the ICPSR (Interuniversity Consortium for Political and Social Research) Summer Program and at the University of Oxford. In organizing my courses, I found that I faced the common problem of providing readings that were appropriate for individuals who were neither statisticians nor econometricians. The problem was partly about finding readings that were targeted at the appropriate level and partly about finding readings that contained appropriate examples, but it was also about finding readings that placed the appropriate emphasis on the goals most important to social scientists. When surveying the literature available, I found myself wishing for a text that placed less emphasis on inductively choosing the model specification and greater emphasis on testing a time series model that is based on theory, and less emphasis on producing good forecasts and more on testing causal relationships. My intention in writing this book is to provide the text that I was seeking for my courses.

This text is intended for social scientists, particularly those who are not econometricians. There are a number of good texts for econometricians already available. However, those in economics who are looking for an alternative perspective might find this book of value. As this text was developed out of a course for graduate students and faculty, they are the target audience. This means that it is assumed that the reader has taken an introductory course in social statistics and linear regression analysis.

This text picks up from linear regression in the cross-sectional setting and ends with error correction models. In doing so, it provides a bridge to the time series techniques necessary for multiple dependent variables, such as vector autoregression. In particular, it provides a bridge to the SAGE QASS (Quantitative Applications in the Social Sciences) text on time series with multiple variables by Patrick Brandt and John Williams (2007). Used on its own, this text provides a good basis for a short course on time series analysis. Used in conjunction with the Brandt and Williams text, this book provides a good basis for a longer course. To facilitate its use both in the classroom and as a resource for the individual researcher, the data and Stata (and R) code for all examples are provided online.

Acknowledgments

I would like to thank Bill Jacoby and the ICPSR Summer Program for providing me the opportunity to teach time series and develop the materials that form the basis of this monograph and for his thoughtful review of the manuscript. I would like to thank John Fox for supporting this project from the beginning to the end. This monograph has also benefited greatly from the comments and input of Paul Kellstedt, Texas A&M University, Jennifer Castle, University of Oxford, and Paul Gustafson, University of British Columbia, as well as the reviewers of the proposal and manuscript, Sarah Mitchell, University of Iowa, and Michael Colaresi, Michigan State University. The monograph is much better for this input, as it is for the assistance from those at SAGE Publications. Finally, I would like to acknowledge the support of my wife, Eline de Rooij, in all endeavors, academic or otherwise.

INTRODUCTION

Time series data are the chronological sequences of observations produced by regularly and repeatedly measuring some characteristic(s) of the same case over time (e.g., aggregate support for the government in a country, the crime rate in a city, etc). Time series analysis is the application of statistical models to time series data. Such analysis is typically done in one of two domains: (1) the frequency domain or (2) the time domain. Time series analysis in the frequency domain involves modelling time series data as a function of an infinite number of cyclical components. Traditionally, this has been done through the use of spectral analysis (Harvey, 1993, chap. 6), but it can also be done using wavelet analysis (Nason, 2008; Torrence & Compo, 1998; Woodward, Gray, & Elliott, 2012, chap. 12). Time series analysis in the time domain is conducted through the analysis of autocorrelations and cross correlations within and between time series data. This book is about time series analysis in the time domain. With rare exceptions (e.g., Aguiar-Conraria, Bagalhães, & Soares, 2012), this is the form of time series analysis used in the social sciences.

Approaches to time series analysis can generally be placed along a spectrum with data-driven approaches on one end and theory-driven approaches on the other. The data-driven end of the spectrum is typified by approaches that are highly inductive. The data are used to determine the appropriate components of the model to be included or excluded. The primary objective of such approaches is to choose the model that best fits the data, often combined with the principle of parsimony (i.e., using the simplest model). The goal of these approaches is often one of using the data to build the optimal model for forecasting, with the principle of optimality including parsimony. At this end of the spectrum are Autoregressive Integrated Moving Average (ARIMA) models selected using the Box-Jenkins approach.

At the other end of the spectrum, at the theory-driven end, the varying approaches rely much more on theory to determine which components to include or exclude. Unlike models derived from data-driven approaches, the components of theory-driven models have a direct interpretation. They represent elements of the theoretically derived data-generating process (often equivalent to the theoretical model) and are commonly called structural models.[1] The goal of such models can also be that of forecasting, but

---

[1] Note, though, that the term structural can be used in a number of ways within the field of time series analysis.

such models are also well suited for testing causal theories. This is because the parameters of the estimated model can be matched up with parameters within the theoretical model. At this end of the spectrum are error correction models, autoregressive distributed lag models, and state space models, among others. Instead of building the most parsimonious model, this approach generally begins with the most general model suggested by theory and then tests what restrictions can be placed on the model. This is called the general-to-specific approach (Spanos, 1986).

The various approaches along the spectrum from data driven to theory driven have varying strengths and weaknesses. The appropriateness of the approach is often a function of the purpose of the analysis. There are three primary uses for time series analysis: (1) to test theory through causal inference, (2) policy analysis, and (3) forecasting. In the first use, the researcher is attempting to use the observed data to test a theory regarding the causal effect of one variable on another. In the second use, the researcher is attempting to determine the effect of a new policy or change in policy on some outcome of interest. In the third use, the researcher is attempting to predict the future value of some variable based on current and past values of the same variable and (possibly) of covariates. Different time series methods are useful for different purposes. For example, simple data-driven approaches are often unbeatable in forecasting but are unsuitable for policy analysis.

This book introduces the reader to approaches from the entire range of the spectrum. However, when choices have to be made for the purposes of exposition, this book tends toward the theory-driven end of the spectrum. This is also the perspective taken when advice regarding model selection is given. This is done as a conscious effort to cater to those analysts interested in estimating the relationships between variables and testing hypotheses about causation and the impact of policy interventions. It is at the theory-driven end of the spectrum that approaches appropriate for these purposes are found.

Two other ways of distinguishing statistical models are worth noting. As in cross-sectional analysis, a distinction can be made between models that contain exclusively observable components and those that include unobservable components, such as latent variables. The distinction can also be made between models that contain a single endogenous (dependent) variable, and therefore a single equation, and those that allow for multiple endogenous variables, and therefore include multiple equations, such as simultaneous equation models.

The time series models covered in this book are of the type that include, almost exclusively, only observable components. The models covered are also single-equation models. However, the assumptions necessary to use a single-equation model, rather than a multiple-equation model, are outlined. Furthermore, the connection between single- and multiple-equation models is made when error correction models are discussed in Chapter 6. Finally, the intent of this book is to provide the reader with the necessary background to understand more advanced texts on multiple-equation models—in particular, Brandt and Williams (2007).

Why Study Time Series Analysis?

Time series analysis is useful for aggregate-level analysis—for example, for analyzing the dynamics in national economic and social statistics, such as crime rates, divorce rates, government social program expenditures, or unemployment rates, or the dynamics in national public opinion, such as popularity of the government or support for a particular policy or military conflict.

The focus of time series analysis on dynamics is useful if the focus of the research is on change. We often want to go beyond the analysis of the relationship between X and Y, to analyze the relationship between the dynamics (changes) in X and the dynamics in Y. This is because social scientists are generally interested in developing and testing causal theories. By analyzing how the dynamics in one variable are related to the dynamics in another variable, time series analysis can be useful for testing the hypotheses derived from such causal theories. One area where this has been put to good use is the examination of the effects of public opinion on government program spending, while accounting for the fact that public opinion is itself affected by government program spending (Soroka & Wlezien, 2010; Stimson, 1999) and media consumption (Johnson & Kellstedt, 2013).

Furthermore, time series analysis is useful for evaluating the effects or consequences of political and social policies. This type of analysis is often called an intervention analysis. For example, what is the effect of introducing seatbelt laws on road traffic fatalities (Harvey & Durbin, 1986)? Or what is the effectiveness of policies designed to fight terrorism (Enders & Sandler, 1993)? In addition to testing the effects of past interventions, time series analysis can also be useful for the purposes of forecasting, for example, forecasting electoral outcomes (inter alia, Erikson & Wlezien, 2012; Fair, 1996) or levels of foreign policy aggressiveness (Moore & Lanoue, 2003).

As we will see, the time series data necessary for these types of analysis can (and most often do) violate some of the most important assumptions of the usual methods used for cross-sectional data analysis. Therefore, a method of analysis appropriate for time series data can be of great value.

Finally, although time series data differ from panel data, an effective understanding of panel data analysis requires an understanding of the time dynamics in the data. Therefore, an understanding of time series analysis is a prerequisite to studying panel data analysis.

This book is intended to be a practical guide to time series models. It will demonstrate the use of, and assumptions underlying, common models of time series data, including static, finite distributed lag, autoregressive distributed lag, moving average (autocorrelated error), differenced data, autoregressive conditional heteroskedasticity (ARCH), generalized autoregressive conditional heteroskedasticity (GARCH), autoregressive moving average (ARMA), autoregressive integrated moving average (ARIMA), and error correction models. These models will be demonstrated with the goal of giving readers the tools necessary to apply them to their own research. An emphasis will be placed on using such models within the fields of public policy, political science, and sociology. In doing so, the book will also provide an introduction to fundamental time series concepts, including autoregression, serial correlation, trending, unit root processes, and the two most important assumptions underlying time series models—stationarity and exogeneity. The monograph will also direct the reader to more advanced readings on topics such as fractional integration and state space models.

The Plan of the Book

In Chapter 1, we define time series analysis and distinguish time series data from other forms of data. You will be introduced to some notation and terminology that will set the framework for the discussion of time series fundamentals in Chapter 2 and two of the examples that will be used throughout the text. We end with a discussion of some of the opportunities and challenges of time series analysis, which will be expanded on in subsequent chapters.

In Chapter 2, you will be introduced to some of the fundamental concepts of time series data and time series analysis: autoregression, autocorrelation, serial correlation, stationarity, exogeneity, weak dependence, trending, seasonality, structural breaks, and stability. These concepts are revisited and extended in later chapters as they become particularly pertinent.

Having covered many of the fundamental concepts of time series in Chapter 2, in Chapter 3 we begin to explore basic models of time series data. This chapter examines the static and finite distributed lag models estimated through ordinary least squares regression. We examine the assumptions required for the estimation of such models to be unbiased. We also examine how to test for and correct violations of key assumptions—in particular, violations of covariance stationarity and no serial correlation. This includes an introduction to models that are estimated using the maximum likelihood approach. Along with the discussion of violations of covariance stationarity, we discuss the challenges of trending, periodicity, and structural breaks.

In Chapter 4, we continue the discussion of modelling time series data with the introduction of dynamic models. These models include the autoregressive distributed lag (ADL), lagged dependent variable (LDV), autoregressive conditional heteroskedasticity (ARCH), and the moving average (MA) models. As the conditions necessary for these dynamic models to meet the assumption of covariance stationarity differ from those of static models, these conditions are discussed.

In Chapter 5, we move on to the ARMA model and the Box-Jenkins approach to building such models. The chapter continues with a discussion of including exogenous regressors in such models for the purposes of estimating the magnitude of their effects and hypothesis testing. This includes a short discussion on transfer functions and intervention analysis. We finish with a discussion of GARCH models.

In Chapter 6, we extend the discussion of ARMA models, begun in Chapter 5, to ARIMA models and introduce error correction models by discussing the concepts of differencing integrated data and cointegration. To that end, the chapter begins with an overview of identifying unit root processes and how one may distinguish these from other forms of trending. Differencing data and the differenced data model are then discussed, and this is used to motivate the ARIMA and error correction models. The reader is also introduced to second, seasonal, and fractional differencing.

ABOUT THE AUTHOR

Mark Pickup is an associate professor in the Department of Political Science at Simon Fraser University. He has taught time series analysis at the Inter-University Consortium for Political and Social Research Summer Training Program since 2010. He is a specialist in comparative politics and political methodology. Substantively, his research primarily falls into three areas: (1) the economy and democratic accountability, (2) polls and electoral outcomes, and (3) conditions of democratic responsiveness. His research focuses on political information, public opinion, the media, election campaigns, and electoral institutions within North American and European countries. His methodological interests concern the analysis of longitudinal data (time series, panel, network, etc.), with a secondary interest in Bayesian analysis. He has published in a variety of leading journals. He holds degrees in chemical physics (BSc) and political science (BA, MA, and PhD). He received his doctoral degree at the University of British Columbia. In addition to his current position at Simon Fraser University, he has been a lecturer at the University of Nottingham and a postdoctoral research fellow at the University of Oxford.

CHAPTER 1: THINKING TIME-SERIALLY

This chapter defines time series analysis and distinguishes time series data from other forms of data. It provides an introduction to some notation and terminology that will set the framework for the discussion of time series fundamentals in Chapter 2 and to two of the examples that will be used throughout the text. It will end with a discussion of some of the opportunities and challenges of time series analysis, which will be expanded on in subsequent chapters.

1.1 Time Series Analysis and Time Series Data

Time series analysis in the social sciences is the application of statistical models to time series data to examine the movement of social science variables over time (e.g., public opinion, government policy, judicial decisions, educational outcomes, socioeconomic measures), allowing analysts to estimate relationships within (over time) and between variables in order to test causal hypotheses, make forecasts about the future, and assess the impact of policy changes.

To clarify exactly what time series data are and are not, it is useful to compare such data to other types of data. For many, the most familiar type of data is cross-sectional data. Typically, cross-sectional data are from a random sample of cases. For example, a variable Y is a collection of observations on randomly selected cases:

$$Y = \{y_1, y_2, y_3, \ldots, y_N\}, \; N = \text{number of cases.} \qquad (1.1.1)$$

Each observation y_i is from a different case, all from the same point in time. If the cases are selected by simple random sampling, each value of y_i is roughly independent of the others. Cases can be a random selection of individuals, countries, firms, and so on.

As an example of cross-sectional data, consider the following cross-sectional data on individual preferences for total government spending in 1976 in Britain (Table 1.1).

Cross-sectional data have one observation for each case. Time series data have a separate observation for each time point, and each observation is for the same case—for example, GDP (gross domestic product) of a country. The time between observations can be years, months, days, hours, and so on. However, as we shall see, the measurements are assumed to be

Table 1.1 Individual Preferences for Total Government Spending in
1976 Britain

Individual	Spending Preference
1	2
2	2
3	1
4	5
5	4
6	5
7	4
8	3
9	5
10	5
11	3
12	4
13	2
14	5
⋮	⋮

NOTE: Individual preferences for total government spending: scored *strongly in favor* (1), *in favor* (2), *neither in favor nor against* (3), *against* (4), or *strongly against* (5) government spending cuts.

(roughly) evenly spaced. A time series variable Y_t is a nonrandom sequence of observations for an individual case ordered over time:

$$Y_t = \{y_1, y_2, y_3, \ldots, y_T\}, \ T = \text{number of time points.} \quad (1.1.2)$$

Again as an example, consider the following time series data on *net* preferences for total government spending in Great Britain each year from 1975 onward (Table 1.2).

Another type of data, which is neither cross-sectional nor purely time series, is panel data. Continuing our previous example, consider the following data on net preferences for total government spending in a selection of countries at three time points: 1986, 1996, and 2006. Panel data can be presented in either stacked or nonstacked format (Table 1.3).

Table 1.2 Net Preferences for Total Government Spending in Great
Britain

Year	Spending Preference
1975	−6.1
1976	−7.9
1977	−9.4
1978	−10.1
1979	−10.7
1980	−11.5
1981	−12.4
1982	−12.6
1983	−12.2
1984	−13.5
1985	−14.8
1986	−14.4
1987	−14.4
1988	−14.3
⋮	⋮

NOTE: Net preferences for total government spending: the average of survey responses, scored strongly in favor (−100), in favor (−50), neither in favor nor against (0), against (+50), or strongly against (+100) government spending cuts. The measure ranges in theory from −100, meaning that all respondents strongly favor spending cuts, to +100, meaning that all respondents oppose spending cuts.

In panel data, we have more than one case. The same set of cases is observed at multiple time points. Typically in the social sciences, we observe more cases than we do time points.

A final type of data is pooled cross-sections. This is not quite panel data as the cases measured at each time point are not the same. For example, our data may be the responses of individuals from Britain to a monthly survey, with a different random sample of individuals each month. This type of data is often collapsed into a time series for the

Table 1.3 Net Preferences for Total Government Spending

Nonstacked Country	1986	1996	2006
Australia	−45.3	−37.3	−12.8
Germany	−51.8	−67.5	−54.9
Great Britain	−14.4	−12.8	−4.5
Hungary	−63.3	−67.2	−63.4
Israel	−73	−70.3	−51.5
Italy	−42.2	−43.5	
Norway	−40.8	−36.4	−30.2
⋮	⋮	⋮	⋮
United States	−53.6	−58.5	−36.1

Stacked Country, Year	Spending Preference
Australia, 1986	−45.3
Australia, 1996	−37.3
Australia, 2006	−12.8
Germany, 1986	−51.8
Germany, 1996	−67.5
Germany, 2006	−54.9
Great Britain, 1986	−7.9
Great Britain, 1996	−12.8
Great Britain, 2006	−4.5
Hungary, 1986	−63.3
Hungary, 2006	−67.2
Hungary, 2006	−63.4
⋮	⋮
United States, 2006	−36.1

purposes of analysis. For example, we could take individual-level total government spending preference data and calculate the net preference for total government spending for the individuals surveyed each month. This gives us a monthly time series of net preference for total government spending in Britain.

Time series, panel, and pooled cross-sectional data are all forms of longitudinal data. Before looking at the analysis of time series data, we need to specify some notation and terminology.

1.2 Time Series Notation and Terminology

One of the major stumbling blocks for students trying to understand time series analysis is the notation used. This is understandable, as reading any quantitative methods literature without knowing the notation is a bit like trying to read a text in a foreign language. Unfortunately, there is no single agreed-on notation, so the notation used in this text may differ from what you read elsewhere, but the notation is consistent throughout this text and the supplementary material. We begin with some basic notation:

X, Y, Z, W	Variables
x, y, z, w	Some single value (element) of the variable (i.e., the value of the variable at some unspecified single time point)
i, j, k, l, s, t	Indices (t is usually reserved to index time)
x_t, y_t, z_t, w_t	A specific value (element) of the variable (i.e., the value of the variable at some specified single time point)
T	Total number of observed time points: $t = 1, 2, 3, \dots, T$

Some notation is specific to time series data. If Y is a time series variable, we often give it the subscript t: Y_t. The subscript t does not indicate that we are referring to a specific value of Y_t. It only indicates that Y is a time series variable. We will use y_t to denote the specific value of Y_t at some time point t. For example, y_1 or $y_{t=1}$ is the specific value of Y_t at the first time point; this is our first observation in the time series.

The Data-Generating Process Versus the Data Model

In the following chapters, we will often describe what we assume to be the data-generating process for the time series data that we are analyzing. This language may be new to many and needs some explanation.

When learning about simple linear regression, the distinction is often made between the *population model* or *true model* and the model estimated from the sample. In this language, the population model describes the process that generates the data from which we sample. For example, for variables X and Y, we may assert the population regression model as follows:

$$Y = \beta_0 + \beta_1 X + \varepsilon,$$

$$\varepsilon \sim \text{NID}\left(0, \sigma_\varepsilon^2\right), X \sim \text{NID}\left(\mu_X, \sigma_X^2\right), E\left(\varepsilon \mid X\right) = 0. \tag{1.2.1}$$

Variable Y is a function of X and ε, which are themselves normally distributed random variables that are unrelated to each other (independent), denoted as NID. In this assertion, we assume that this is the process that generates the set of data that is our sample. Our sample represents N draws from this stochastic (containing a random component) process— specifically N random draws from X and ε, which then determine Y.[1] From the sample, we can specify the sample regression function as follows:

$$Y = \beta_0 + \beta_1 X + \varepsilon,$$

$$Y = \{y_1, y_2, \ldots, y_N\}, X = \{x_1, x_2, \ldots, x_N\}. \tag{1.2.2}$$

This model, which we will estimate using our sample data (e.g., using ordinary least squares [OLS] estimation), is called the *data model* (sometimes called the *empirical model*). We indicate the model from a particular estimation using the "hat" notation:

$$Y = \hat{\beta}_0 + \hat{\beta}_1 X + \hat{\varepsilon}. \tag{1.2.3}$$

The data-generating process often is described as a stochastic function that could produce an infinite number of possible outcomes. Our data are N possible draws from this function. In many cross-sectional contexts, it is easy enough to think of the sample data as a random draw of N cases from the *very large* population of cases available for observation. The randomly selected cases provides us with our sample values of Y, X, and ε. In thinking about our data this way, it isn't really necessary

---

[1] Note that X may also contain nonstochastic elements.

to use the language of a data-generating process—although it may still be useful.

In the context of time series data and often in the context of cross-sectional data, our sample data are not a random draw from a population of cases available for observation.[2] Consequently, the language of a data-generating process becomes not just useful but necessary. In the time series context, the data-generating process is again described as a stochastic function that could produce an infinite number of possible outcomes. For example,

$$y_t = \beta_0 + \beta_1 x_t + \varepsilon_t \text{ for } t = 1, 2, \ldots, T$$

$$\varepsilon_t \sim \text{NID}\left(0, \sigma_\varepsilon^2\right), \ x_t \sim \text{NID}\left(\mu_X, \sigma_X^2\right), E\left(\varepsilon_t \mid x_t\right) = 0. \quad (1.2.4)$$

However, our data are a single draw from this data-generating process, in that we only ever draw one value of x_t and one value of ε_t for each time point $t = 1, 2, \ldots, T$. These then determine the values of y_t observed. The single draw of x_t and ε_t for each time point $t = 1, 2, \ldots, T$ is commonly called a single *realization* of the data-generating process. In time series analysis, we are not in a position to go back and resample different values of x_t and ε_t for a particular time point. Talking about the population of cases we could have observed at a single time point is meaningless, and so we instead talk about the stochastic process that generates the single value at that time point.

When describing fundamental time series concepts, we will often define the data-generating process that corresponds with each of the concepts. When describing the application of time series analysis, we will discuss the consequences of different data-generating processes for the model we estimate from our data (the data model). It will become evident that it is not always necessary for the estimated data model to contain all of the elements of the assumed function defining the data-generating process. We may assume that the data-generating process is as follows:

$$y_t = \beta_0 + \beta_1 x_t + \beta_2 z_t + \varepsilon_t \text{ for } t = 1, 2, \ldots, T$$

$$\varepsilon_t \sim \text{NID}\left(0, \sigma_\varepsilon^2\right), \ x_t \sim \text{NID}\left(\mu_X, \sigma_X^2\right), \ z_t \sim \text{NID}\left(\mu_z, \sigma_z^2\right), \ E\left(\varepsilon_t \mid x_t, z_t\right) = 0. \quad (1.2.5)$$

---

[2] For example, when the cases in the cross-sectional data are European countries, we never really have a random draw of these countries.

At the same time, we might explore the consequences of this data-generating process for an OLS estimation of a data model of the following form:

$$y_t = \beta_0 + \beta_1 x_t + \varepsilon_t, \text{ for } t = 1, 2, \ldots, T \qquad (1.2.6)$$

It is usually the case that we believe the data-generating process is more complicated than the data model estimated. It is not uncommon to be aware that y_t is, in part, determined by z_t without having any direct measure of z_t. The data-generating process can be interpreted as the unobserved reality that we are trying to reveal with our analysis. However, much analysis is focused on trying to reveal only a part of the data-generating process, while guarding against the possibility that the larger reality might lead us to reach false conclusions.

In time series analysis, as in other forms of statistical analysis, there is often an iterative process between the data-generating process assumed and the data model estimated (Hendry, 2003, chap. 1). Typically, we begin by stating the assumed data-generating process. This informs the data model that we then estimate. The results from the estimated data model may provide confirmation that our data-generating assumptions are correct or may contradict those assumptions. Accordingly, we may adjust our assumptions regarding the data-generating process and estimate a new data model. This process continues until we feel that there has been a convergence between our assumptions regarding the data-generating process and the data model estimated from our sample data.

The Lag of a Time Series Variable

Continuing on with time series notation, we use the notation y_{t-1} to indicate the specific value of Y at the time point just previous to t. We call this the *lag* of y_t. Say our observations are weekly; then, y_{t-1} at $t = 5$ (Week 5) is equal to $y_{5-1} = y_4$. This is the value of Y_t in Week 4. For clarity, this is sometimes called the *first* lag. Similarly, y_{t-2} is called the *second* lag. This can, of course, be applied to any variable (e.g., X_t or Z_t).

To "lag" a variable is to create a new variable where the value of the variable at a given time point t is replaced by the value of the variable from the previous time point, $t - 1$. Consider the following data (Table 1.4) containing the variable "social program spending preference" (measured on the same scale as the "total government spending net preference" variable in the previous example) in the third column. The lagged value of social program spending preference would look as it does in the fourth column. If we are modelling social program spending preference

Table 1.4 Social Program Spending Preference—Lags, Leads, and First Differences

Year	Unemployment	Social Program Spending Preference	Social Program Spending Preference —First Lag	Social Program Spending Preference —First Difference	Social Program Spending Preference —First Lead
1988	7.71	27.45	—	—	32.32
1989	7.59	32.32	27.45	4.87	25.18
1990	8.84	25.18	32.32	−7.15	22.39
1991	10.46	22.39	25.18	−2.79	20.26
1992	11.35	20.26	22.39	−2.13	15.53
1993	11.38	15.53	20.26	−4.74	20.63
1994	10.08	20.63	15.53	5.10	23.80
1995	9.51	23.80	20.63	3.17	33.41
1996	9.71	33.41	23.80	9.62	40.13
1997	8.95	40.13	33.41	6.72	45.08
1998	8.15	45.08	40.13	4.95	46.69
1999	7.30	46.69	45.08	1.61	44.89
2000	6.90	44.89	46.69	−1.80	—

Note that in practice you lose a time point when lagging a variable. In discussions of the theory of time series analysis, this is sometimes overlooked, but it will be important to keep in mind when we discuss the practical application of time series analysis.

and we include a lag of this variable as an explanatory variable, this variable is often called a "lagged dependent variable," as it is just that—the lagged value of the dependent variable. More will be discussed on this in Chapters 2 through 4.

The Lead of a Time Series Variable

To "lead" a variable is to create a new variable where the value of that variable at a given time point *t* is replaced by the value of the variable from the *next* time point, $t + 1$. The lead value of "social program spending preference" would appear as it does in the last column of Table 1.4. The notation for the lead of y_t is y_{t+1}. It is rare that such a variable would be included as an explanatory variable in a model. It is unlikely that we would expect the future value of the dependent variable to be an explanatory variable, but in Chapter 6 we will see an example of its use.

The Difference of a Time Series Variable

To first difference a time series variable is to create a new variable where the value of that variable at a given time point t is equal to the value of the original variable minus the value of the first lag of the variable. The second-last column of Table 1.4 contains such a first difference of social program spending preference. The 1989 value of 4.87 is the result of subtracting the lag of social program spending preference (27.45) from the original 1989 value of social program spending preference (32.32). The notation for the first difference of y_t is Δy_t ——the Greek letter *delta* is often used to indicate change or difference. The first difference of a variable can be interpreted as the change in this variable since the previous time point. The change in social spending preference from 1988 to 1989 is 4.87 (an increase).

Getting a grasp on notation and terminology is one of the greatest hurdles to understanding time series analysis. This text will build on the notation and terminology outlined in this chapter as needed, but what has been presented so far provides the framework necessary for the discussion of time series fundamentals in Chapter 2. Before we move on to those fundamentals, let us discuss some of the key opportunities and challenges presented by time series analysis.

1.3 Opportunities and Challenges With Time Series Data

To explore the problems created by time series data for the methods of analysis generally used with cross-sectional data, let us consider the following example. For this example, we will analyze data collected for the purpose of testing the thermostatic model of public responsiveness and policy representation as developed by Soroka and Wlezien (2010).

The thermostatic model is actually two models that describe how (1) public demands for increases or decreases in policy spending respond to current levels of government spending and (2) government changes in policy spending respond to public demands for increases/decreases. The following examines the first of these, called a public responsiveness model. The unit of time for this model is the fiscal year—there is one observation per fiscal year. The public responsiveness model is as follows:

$$R_t = \beta_0 + \beta_1 P_t + \beta_2 W_t + \varepsilon_t \qquad (1.3.1)$$

where R_t is the public's relative preference for policy spending in a given year—that is, the difference between the public's preferred level of policy spending and the level that it actually gets.[3] P_t is the actual level of policy spending in a year. W_t represents other, exogenous effects on the public's relative preferences—this could include more than one variable. For our current purposes, let us just regress the public's relative preference for policy spending on the actual level of policy spending (in millions of Canadian dollars). We do this using yearly data on the Canadian public's relative preference and government policy spending for social welfare payments from 1988 to 2003. Table 1.5 contains the results from an OLS estimation.

In addition to estimating the intercept and slope coefficients, β_0 and β_1, and calculating the corresponding t statistics, we usually estimate a goodness-of-fit statistic (e.g., R^2).

Recall that we typically use a t test to test the statistical significance of the regression coefficients. Looking at Table 1.5, if we were using a 0.05 significance level, we would conclude that we could not reject the null hypothesis that the slope coefficient for program spending is 0. Therefore, we could not reject the null hypothesis of no effect for program spending on relative spending preference. If we were using a 0.10 significance level, we would conclude that there is a statistically significant and positive relationship between government social welfare spending and the public's relative preference. For each additional billion dollars spent on

Table 1.5 Canadian Public's Response to Government Spending

Preference	Coefficient	Standard Error	t Statistic	P Value
Program spending	0.32	0.16	1.98	0.067
Constant	−25.37	30.03	−0.84	0.412

NOTE: $R^2 = 0.22$, $T = 16$; T = number of time points.

---

[3] This is operationalized using the measure described in Table 1.2.

social welfare programs, the public's relative preference increases by 0.32 on the (−100, 100) scale. This suggests that greater spending leads to an increase in the demand for spending! Neither of these is what Soroka and Wlezien (2010) predicted, but this is also not the model they used—and for good reason.

The preceding analysis included a number of assumptions. One of the most important of these assumptions is that the observations are independent. Usually, the cases in a cross-sectional data set are assumed to have been selected randomly, and therefore, the value of any case for a given variable will be independent from the value of any other case for the same variable. In time series data, we have measures of the same variable for the same single case at different time points. Therefore, time series data, such as those used here, are usually not independent, especially if the sampling time interval is small. Observations close together are often more alike than those far apart. For example, public opinion today is more closely related to public opinion yesterday than it is to public opinion last year. In our current example, this is public opinion regarding government spending levels. Addressing the potential for time series data to violate the assumption of independence motivates many of the analytical approaches discussed in this text. For now, let us consider some of the consequences of this violation.

If not accounted for in our analysis, one of the problems the violation of independence can lead to is a problem called serial-correlated errors. This is the problem of correlation across the estimated errors in our data model. We discuss this further in Chapters 2, 3, and 5. For now, consider the possibility that the effects captured by the error term at one time point may be correlated with the effects captured by the error term at the next time point. Relating back to our example, this could happen if an event captured by the error term one month increases the public's wish for spending and, at least, part of this effect remains and is captured by the error a month later.

A second potential problem is as follows. A source of the lack of independence may be that public opinion today is partly explained by public opinion yesterday. If we are modelling public opinion on a daily basis and public opinion yesterday is a predictor of public opinion today, this predictor should be included in the model. This predictor is the lag of the dependent variable discussed earlier. The substantive interpretation of including such a variable in a model will be discussed in Chapters 2 and 3. For now, it is sufficient to note that not including it under these circumstances can lead to a violation of another important assumption of OLS regression. This is the zero conditional mean assumption. Let us state this assumption in terms of our current example:

$$E(\varepsilon_t | P_t) = 0. \qquad (1.3.2)$$

This implies that the (conditional) mean of the error term is independent from the explanatory variable(s). What if the past year's relative policy spending preference is a predictor of the current year's relative policy spending preference (R_t), as suggested above? If it is also a predictor of our explanatory variable—current policy spending (P_t)—Assumption (1.3.2) will be violated. This may be the case if government spending levels are a function of the public's relative preference in the previous fiscal period. This is very similar to the omitted-variable bias problem familiar to those who have studied cross-sectional data analysis. It is called an endogeneity problem and is discussed further in Chapter 2.

Those who have studied cross-sectional data analysis will be familiar with the problem of omitted-variable bias. This is a particularly pernicious problem as there is no direct way of testing for omitted-variable bias except for including the variable(s) that is(are) suspected to be omitted. The difficulty that often arises is that a variable that is commonly suspected to be omitted is the past value of the dependent variable. In the figure below, we may be interested in the effect of x_t on y_t, but we are concerned about the effect of y_{t-1} on both x_t and y_t in the data-generating process. In other words, we are concerned that the omission of y_{t-1} from the data model might produce an omitted-variable bias.

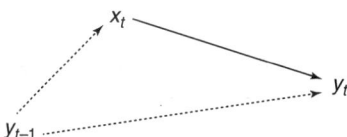

This concern is actually relatively common. Consider this problem in the context of the current example. We regress public attitudes toward government spending levels in a given fiscal period on actual government spending in the same fiscal period:

An omitted-variable bias will occur if public attitudes toward spending in this fiscal period are correlated with public attitudes in the previous fiscal period and if those attitudes in the previous fiscal period affect government spending this fiscal period.

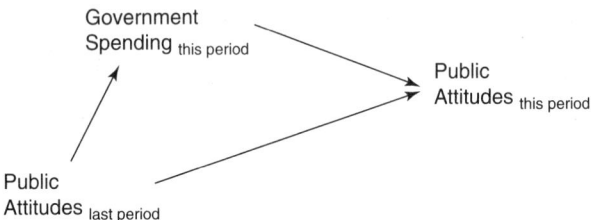

This is one of the more common violations of the zero conditional mean assumption, known as the exogeneity assumption. This assumption is discussed further in Chapter 2. As we will see, time series analysis allows us to test and correct for the problem by including the past value of the dependent variable directly in the data model. This is one way in which time series data provide us with an opportunity we do not have with cross-sectional data. We will explore other opportunities throughout this text.

The analysis of time series data also introduces additional problems. For example, another potential violation of the zero conditional mean assumption is that both x_t and y_t are trending. A variable trends if, in addition to other dynamics and random variation, it increases or decreases by a constant magnitude each time period. This could occur for a number of reasons, such as if social welfare programs have steadily become more expensive to provide over time and the public's expectations regarding the provision of those programs have also increased steadily. Let us visually examine the variables: the public's relative preference for social welfare spending and actual social welfare spending (Figure 1.1).

Clearly, both variables are trending upward. If two series are trending together, we will probably estimate a strong correlation between the two, but we can't assume that the relation is causal. An alternate possibility is that it is a spurious result produced by the fact that both variables are a function of time. The difficulties with and approaches to trending are discussed further in Chapters 2, 3, and 6.

Generally, if we are interested in the effect of x_t on y_t, we need to be concerned if both x_t and y_t appear to have data-generating processes that are a function of time. If both x_t and y_t trend and this is not accounted for in the data model, our estimation of the effect of x_t on y_t will be subject to a spurious correlation, akin to omitted-variable bias.

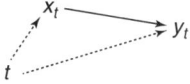

As another concrete example, it is quite common for a new government's popularity to begin trending downward after an initial honeymoon period. This could be driven by any number of things: The new government's popularity was artificially high due to the positive coverage from the election win; once the government has to start making decisions it inevitably upsets some supporters; and so on. As a consequence, any other variable that trends downward or upward during the same period will correlate significantly with government popularity even if it is not related to it in any way. This will hold for other functions of time as well, as we will discuss in Chapter 3.

Trending also violates an assumption that is unique to longitudinal (e.g., time series and panel) data. This is the assumption of stationarity. The nature of the data-generating process of time series data discussed earlier in

Figure 1.1 Canadian Public's Relative Preference and Actual Government Spending

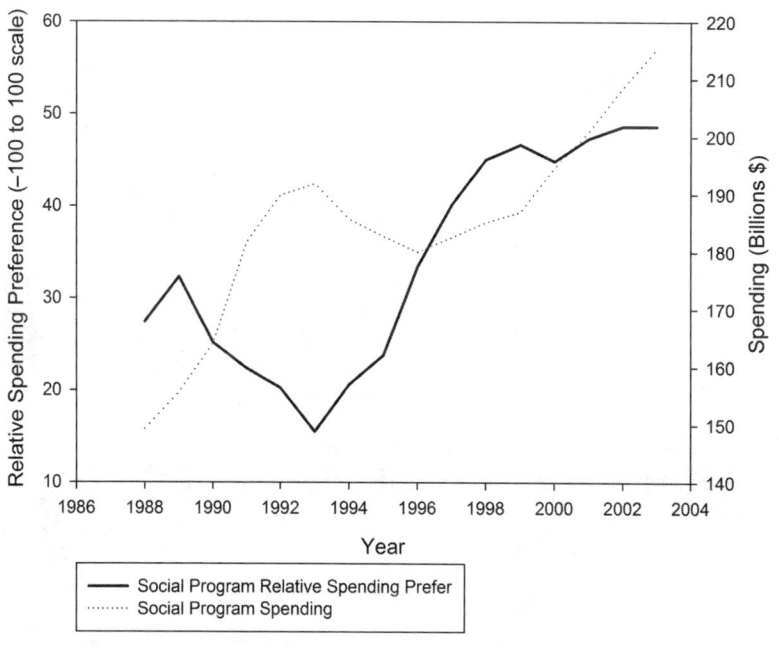

this chapter will require us to make this complex assumption. This is discussed further in Chapter 2.

A final challenge presented by time series data that we have not yet touched on can be illustrated by another example. We may be interested in how approval ratings for the German government translate into vote intention for the government over the period from 1982 to 1998 (Figure 1.2). Over this period, our data are for the government of West Germany prior to January 1987 and for the government of the unified Germany subsequently. If we plot our approval and vote intention time series, we will note something important.

After the reunification of Germany, vote intention becomes somewhat more volatile and approval even more so. This is likely due to the weakening of partisan identification in the latter period (Pickup, 2010). What we are seeing is a structural break in the variances of both time series. This

Figure 1.2 German Government's Approval and Vote Intention

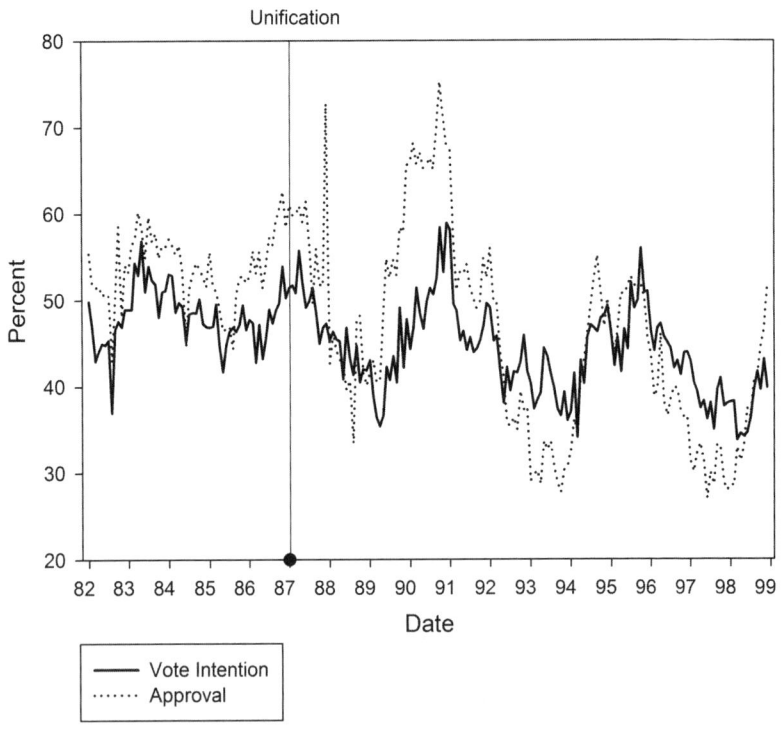

structural break will present a problem for any model that assumes that the variances in the time series are constant across the period of analysis. Structural breaks can also occur in the means of time series and the covariances between time series. In fact, it is quite likely that the covariance between approval and vote intention also has a structural break in January 1987. We will need to account for such breaks in our time series models. As we will see in Chapter 5, such structural breaks will sometimes be a nuisance to be dealt with but at other times they will be of interest in and of themselves.

Summary

In this chapter, you have been introduced to some of the basic notation and terminology of time series analysis. You have also been given a taste of the opportunities and challenges presented by the analysis of time series data. In the following chapters, we will explore these and other opportunities and challenges further. We will learn how to address the challenges and take advantage of the opportunities. In the next chapter, we continue our introduction to time series analysis by surveying the fundamental concepts of time series data and analysis.

CHAPTER 2: FUNDAMENTAL CONCEPTS IN TIME SERIES ANALYSIS

In this chapter, you will be introduced to some of the fundamental concepts of time series data and time series analysis: autoregression, autocorrelation, serial correlation, stationarity, exogeneity, weak dependence, trending, seasonality, structural breaks, and stability. Some of these topics we touched on in Chapter 1. Other topics will be new to you. We will revisit and extend each of these concepts in later chapters, as they become particularly pertinent.

2.1 Autoregression

If Y_t is a time series process, we say that it is an autoregressive process of order 1 if Y_t is a function of its previous value (the lag of Y_t, denoted Y_{t-1}) and a stochastic error (ε_t):

$$y_t = \alpha_1 y_{t-1} + \varepsilon_t, \tag{2.1.1}$$

with $\varepsilon_t \sim \text{NID}(0, \sigma_\varepsilon^2)$—that is, the ε_t values are independent and normally distributed with mean 0 and variance σ_ε^2. We use the following notation for an autoregressive process of order 1: AR(1). Without a constant, Equation 2.1.1 assumes that the long-run equilibrium for the process (if one exists) is 0. The equilibrium can be thought of as the expected value (average) of y_t over a very long period of time. This is discussed further in the next chapter, but for now it suffices to say that this assumption can be relaxed by including a constant: $y_t = \alpha_0 + \alpha_1 y_{t-1} + \varepsilon_t$.

Equation 2.1.1 describes the data-generating process for an autoregressive process. Such a time series process is modelled by including lags (or just a single lag) of the dependent variable on the right-hand side (as explanatory/independent variables). The unknown parameters α_1 and σ_ε^2 are estimated.

The subjective interpretation of a process that is autoregressive is that it has "memory." Assume that $|\alpha_1| < 1$. This is an assumption to which we will return. If something occurs to move the process out of its long-run equilibrium (an intervention), it will, in the absence of any other intervention, return to its long-run equilibrium. However, it will not return immediately. In the periods following the intervention, the process will exhibit the decaying effect of the intervention until the process returns to equilibrium.

In the public responsiveness model of Chapter 1, we noted that it might be prudent to model relative preference as a function of past relative preference in order to avoid a violation of the zero conditional mean assumption. The model based on Equation 2.1.1 is just the sort of model we might use to do this.

2.2 Autocorrelation

A concept related to autoregression is autocorrelation. We define the correlation between the elements (Y_t) of a time series Y_t and the same time series lagged once, Y_{t-1}, as follows:

$$\text{Corr}\left(y_t, y_{t-1}\right) \equiv \rho \equiv \frac{E\left(\left(y_t - \mu_y\right)\left(y_{t-1} - \mu_y\right)\right)}{E\left(y_t - \mu_y\right)^2}. \tag{2.2.1}$$

Note: $E(\ )$ denotes the expected value $\mu_y = E(y_t)$ and μ_y denotes the theoretical mean of y_t. This is the covariance between y_t and its own lag y_{t-1} divided by the variance of y_t. This is called autocorrelation. The covariance of the element of a variable with a lag of itself is called autocovariance, although if its meaning is clear from the context, it will often simply be called covariance.

Under the assumption of stationarity, discussed below in Section 2.4 we can estimate the autocorrelation as follows:

$$\hat{\rho} = \frac{\sum_{t=2}^{T}\left(y_t - \bar{y}\right)\left(y_{t-1} - \bar{y}\right)}{\sum_{t=2}^{T}\left(y_t - \bar{y}\right)^2}. \tag{2.2.2}$$

Note: $\bar{y}$ denotes the mean of the observed values of Y_t. Also, note that we begin the summation at $t - 2$ to reflect the fact that in practice we lose the first data point when we lag a variable. The relationship between autocorrelation and autoregression can be seen in the fact that when we model Y_t as an AR(1) process, as depicted in Equation 2.1.1, the estimate of the α_1 coefficient is the estimated autocorrelation $\hat{\rho}$.

This autocorrelation is called first-order autocorrelation, as it is the correlation between a time series and the *first lag* of itself. An extension is higher-order autocorrelation. For example, the elements of a time series Y_t

may be correlated with the elements of the second lag of itself, Y_{t-2}. We would call this second-order autocorrelation and define it as follows:

$$\rho_2 \equiv \frac{E\left(\left(y_t - \mu_y\right)\left(y_{t-2} - \mu_y\right)\right)}{E\left(y_t - \mu_y\right)^2}. \tag{2.2.3}$$

Generalizing this to lag s,

$$\rho_s \equiv \frac{E\left(\left(y_t - \mu_y\right)\left(y_{t-s} - \mu_y\right)\right)}{E\left(y_t - \mu_y\right)^2}. \tag{2.2.4}$$

These are unknown for the true data-generating process, but they can be estimated from the sample data, assuming that the series is stationary:

$$\hat{\rho}_s = \frac{\sum_{t=s+1}^{T}\left(y_t - \bar{y}\right)\left(y_{t-s} - \bar{y}\right)}{\sum_{t=1}^{T}\left(y_t - \bar{y}\right)^2}. \tag{2.2.5}$$

As an example, consider the second example described in Chapter 1. We can examine the monthly vote intention data (the percentage of the population indicating that they would vote for the governing party in an election) for the German government of Helmut Kohl from October 1982 till September 1998 (Figure 2.1). The first-order autocorrelation is estimated to be 0.62, and the second-order autocorrelation is estimated to be 0.57. We can plot the first-, second-, and higher-order autocorrelations in a single figure, as in Figure 2.2. This is called an autocorrelation function and will be discussed further in Chapter 5.

We can also conduct a test of whether there is any evidence that the time series data depart from a white noise process. This process is defined as follows:

$$y_t = \varepsilon_t, \varepsilon_t \sim \text{NID}(0, \sigma_\varepsilon^2). \tag{2.2.6}$$

A white noise process (Equation 2.2.6) is equal to a stochastic error with constant variance. Such a process has no autocorrelation of any order.

22

Figure 2.1 West German Government's Popularity, 1982 to 1998

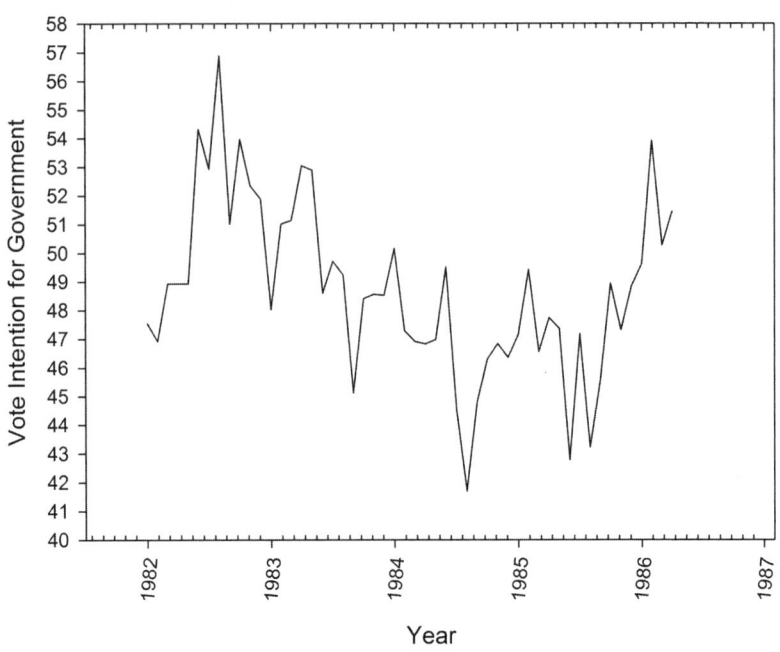

As with the AR(1) process, this can be given a nonzero mean by including a constant, β_0. One of the most common tests for a white noise process is the Portmanteau (Q) test (Ljung & Box, 1978):

$$Q = T(T+2)\sum_{s=1}^{p}\frac{\hat{\rho}_s^2}{T-s}. \qquad (2.2.7)$$

This tests that the error autocorrelations are jointly zero, based on the first p autocorrelations:

$$H_0: \rho_1, \rho_2, \ldots, \rho_p = 0.$$

The choice of p is somewhat arbitrary. High p will capture autocorrelations at high lags but reduces the power of the test. Statistical packages have a default setting. For example, Stata uses either a p of 40 or ($T/2$) − 2,

Figure 2.2 Autocorrelation Function for Vote Intention

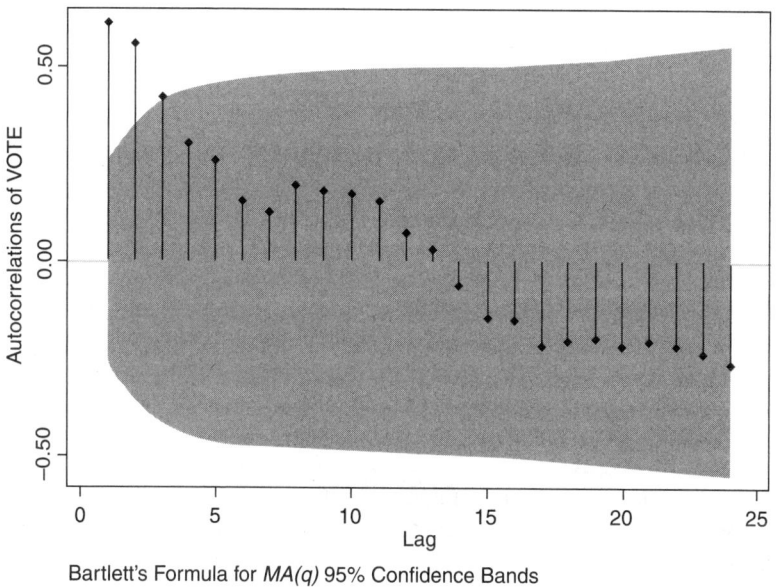

Bartlett's Formula for *MA(q)* 95% Confidence Bands

whichever is smallest. The Q-test statistic is chi-squared distributed with p degrees of freedom, so the P value can be calculated for the purpose of hypothesis testing. The null hypothesis can be interpreted as "The process is white noise."

In Chapter 3, we will look at additional ways of testing whether a process is autocorrelated. Such tests will be useful as many of our time series models depend on the assumption that the error term is a white noise process—that is, it is not autocorrelated. In Chapter 1, we discussed how correlated errors will result from a violation of the assumption of independence of observations. We call this serial correlation. We will want to test the assumption of no serial correlation by applying such white noise tests to the residuals from our models.

2.3 Serial Correlation

If the error term in a statistical model contains autocorrelation (see Section 2.2), this is often referred to as serial correlation. For example, consider the time series process Y_t modelled as a function of a constant, a single independent variable X_t, and an error term:

$$y_t = \beta_0 + \beta_1 x_t + \varepsilon_t. \tag{2.3.1}$$

An example of serial correlation would be if ε_t was correlated with a lag of itself. That is,

$$\mathrm{Corr}(\varepsilon_t, \varepsilon_{t-h} | X) \neq 0 . \tag{2.3.2}$$

The conditioning on X in Equation 2.10 refers to the fact that it is the correlation in the errors of the model that contain X_t that is relevant to discussions of serial correlation. It may be the case that including an additional independent variable in a data model might result in errors that are not serially correlated. The model we ran for the public responsiveness example had the same form as Equation 2.3.1. Including a lagged-dependent variable, as we discussed, might not just avoid a violation of the zero conditional mean assumption; it might also ensure errors that are not serially correlated. We will discuss other forms of serial correlation and methods to address them in Chapter 3 and models that include a lagged-dependent variable in Chapter 4.

2.4 Stationary Stochastic Process

As is common with cross-sectional variables, time series variables contain a stochastic component. For example, the data-generating process for Y_t may be a function of a constant, zero or more independent variables, and a random error:

$$y_t = \alpha_0 + \beta_1 x_t \ldots + \varepsilon_t. \tag{2.4.1}$$

The stochastic error component means that our particular observed values of Y_t are only one possible realization of this time series process. The particular values we observe will depend on one particular realization (draw) of the error term for each time point (as well as the observed values of x_t): $\{\varepsilon_1, \varepsilon_t, \ldots, \varepsilon_T\}$. As this component is stochastic, another draw could have been observed. Each random draw of the set of values $\{\varepsilon_1, \varepsilon_t, \ldots, \varepsilon_T\}$ results in a different realization of Y_t. Of course, we can only ever witness one realization of Y_t. For each point in time, we only ever have one observation. This is distinct from cross-sectional data, as it is (usually) easy to conceptualize drawing additional observations for a single point in time from the population of cases.

In the case of cross-sectional data, we use regression analysis to predict $E(Y)$, based on the average value of Y across the cases in our sample (this is usually done conditioning on independent variables: $E(Y|X)$). The equivalent in time series analysis would be predicting $E(y_t)$ based on the average value of y_t for all cases at time t.

If we could, we would take a random sample of realizations of Y_t and estimate $E(y_t)$ for a particular time point t from the sample of values for y_t at that time point. But we only have one value of y_t at time point t. Therefore, we average the values across all time points ($t = 1, \ldots, T$) of a single realization of Y_t to estimate $E(y_t)$.

This presents a problem. For this to be valid, we must assume that the expected value of Y_t across time converges (as the number of time points increases) on the expected value of y_t for any single time point t. This then implies that the expected value across time of a single realization of Y_t converges on the expected value of y_t for all time points t.

$$\text{As } T \to \infty: E(Y_t) \to E(y_t) \text{ for all } t, t = 1, \ldots, T \qquad (2.4.2)$$

Note: The left expectation is the mean of the values of a single realization of Y_t across all time points, while the right expectation is the mean of all possible realizations of y_t at any specific t. This implies that $E(y_t)$, and therefore $E(Y_t)$, must be constant over time. This is also true for estimating the variance and autocovariances of the y_ts from the variance and autocovariances of our single realization of Y_t.

This brings us to the idea of stationarity. A stochastic time series process Y_t is stationary if the distributions and joint distributions of the y_ts from $t - 1$ to $t = h$ are the same as they are from $t = 1 + k$ to $t = h + k$ for every and any $k \geq 1$.

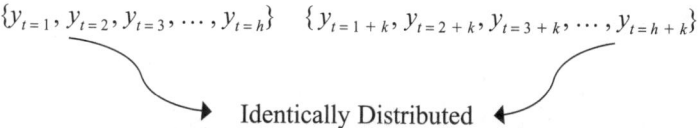

$$\{y_{t=1}, y_{t=2}, y_{t=3}, \ldots, y_{t=h}\} \quad \{y_{t=1+k}, y_{t=2+k}, y_{t=3+k}, \ldots, y_{t=h+k}\}$$

Identically Distributed

Thus, stationarity implies that the y_ts are identically distributed over time. Put another way, all marginal and joint distributions of the process are invariant to time. Therefore, we can estimate the parameters of the distributions of the y_ts from the distribution of the single realization of Y_t.

Covariance-Stationary Process

Usually, for the type of time series analysis covered in this book we shall assume a weak form of stationarity. This is called *covariance* stationarity. A stochastic process is covariance stationary if $E(y_t)$ is constant, $\text{Var}(y_t)$ is constant, and for $k \geq 1$, $\text{Cov}(y_{t=k}, y_{t=h+k})$ depends only on h and not on k (Greene, 2003, p. 612). The assumption regarding the covariances of y_t means that the covariance between y_t and y_{t-1} may differ from (1) y_t and y_{t-2} but not from (2) y_{t-1} and y_{t-2}. The difference is that (1) is the covariance between y_t and its second lag, which can differ from the covariance between y_t and its first lag. However, (2) is the covariance between y_t and its first lag at a different point in time. It is assumed that this makes no difference.

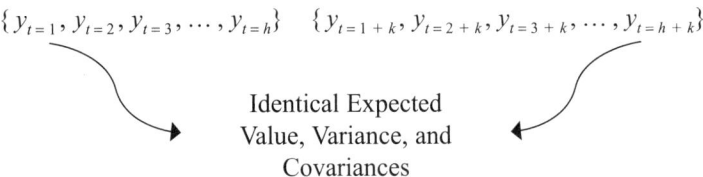

$$\{y_{t=1}, y_{t=2}, y_{t=3}, \cdots, y_{t=h}\} \quad \{y_{t=1+k}, y_{t=2+k}, y_{t=3+k}, \cdots, y_{t=h+k}\}$$

Identical Expected
Value, Variance, and
Covariances

If Y_t is a covariance-stationary process, the expected value (mean), variance, and covariances of the y_ts can be approximated or estimated from a single realization of Y_t. For the mean,[1]

$$\bar{y}_t = \frac{\sum_{t=1}^{T} y_t}{T}. \tag{2.4.3}$$

For the variance,

$$\widehat{\text{Var}}(y_t) = \frac{\sum_{t=1}^{T} (y_t - \bar{y}_t)^2}{T - 1}. \tag{2.4.4}$$

For the h-order (auto)covariance,

$$\widehat{\text{Cov}}(y_t, y_{t-h}) = \frac{\sum_{t=1}^{T} (y_t - \bar{y}_t)(y_{t-h} - \bar{y}_t)}{T - 1}. \tag{2.4.5}$$

However, estimating the h-order covariance requires lag h of y_t. This will result in the loss of h data points, as described in Chapter 1—each time you lag an observed variable, you lose one data point. So unless we are in a position to collect additional data, for the h time points before our first observation, the estimate will actually be based on the following:

---

[1] We could also denote this as $\bar{Y}_t$, but we are assuming that this is the same as $\bar{y}_t$ for all t.

$$\widehat{\text{Cov}}\left(y_{t,}y_{t-h}\right) = \frac{\sum_{t=1+h}^{T}\left(y_t - \overline{y}_t\right)\left(y_{t-h} - \overline{y}_t\right)}{T - h - 1}. \qquad (2.4.6)$$

Instead of the covariances, we often examine the autocorrelations. These are just the covariances divided by the time series variance. We can get an idea of what type of time series will and will not be covariance stationary by considering the weekly volume of Google searches for "The Beatles" and comparing this with the weekly volume of Google searches for "Lady Gaga." Weekly data for these Google searches from the beginning of 2008 until the end of 2011 are displayed in Figure 2.3. These data are the weekly volume of Google searches for "The Beatles" and "Lady Gaga" relative to the total number of searches on Google. Each series is scaled to the mean value for the 2004–2011 period: A value of 1 indicates that the volume is equal to this mean; a value of 2 reflects a volume of twice this mean.

If we like, we can split the 2008–2011 period into two 2-year periods: 2008–2009 and 2010–2011. If the relative search volume is covariance stationary, then the mean, variance, and covariances (or autocorrelations) for the search volume time series should be the same in the first 2-year period as in the second 2-year period. Importantly, the assumption is that this is true for the data-generating process. However, we would also expect this to be approximately true in the observed data.

Looking at the plotted data, we would not expect the Lady Gaga search volume to be covariance stationary. The average volume is much greater in the second period relative to the first. For the Beatles search, it seems possible that the mean volumes in the first and second periods are the same. However, there is some volatility in the first period of the Beatles search volume, which is not seen in the second period—this might mean that the variances differ. The covariance stationarity assumption is about the mean, variance, and autocovariances/autocorrelations of the data-generating process, but we can look at estimates of these from our data to check the plausibility of the stationarity assumption.

For the Beatles search, the estimated mean volume is 0.93 for the 2008–2009 period and 0.88 for the 2010–2011 period. These two means are not exactly the same, but given that they are estimates, it is possible that they both come from the same data-generating process with the same mean. For the Lady Gaga search, the estimated mean volume is 94 for the 2008–2009 period and 220 for the 2010–2011 period. Clearly, the mean volume for the Lady Gaga search is much higher in the second of these two periods.

For the Beatles search volume, the estimated variance is 0.08 for the 2008–2009 period and 0.02 for the 2010–2011 period. For the Lady

Figure 2.3   Weekly Volume of Google searches for "The Beatles" and "Lady Gaga"

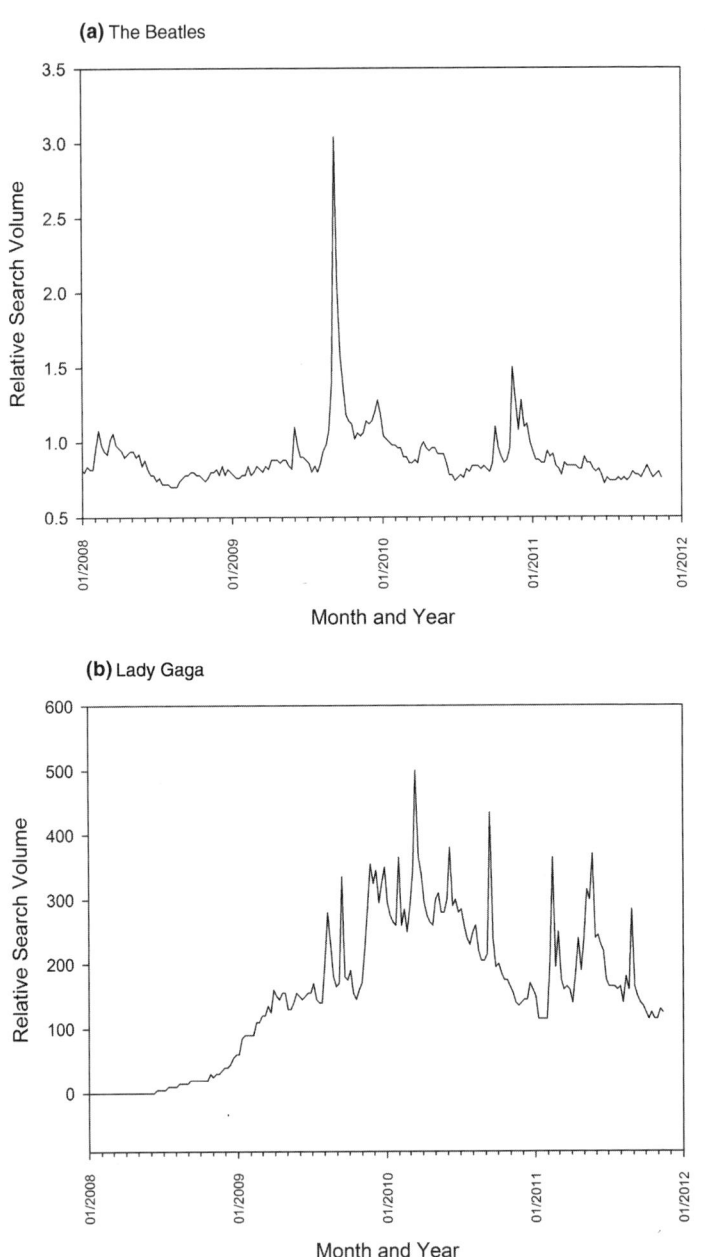

Gaga search volume, the estimated variance is 9,683 for the 2008–2009 period and 6,379 for the 2010–2011 period. The proportionate difference is actually bigger for the Beatles search volume. This is primarily due to the spike in searches in the first 2 weeks of September 2009, because digitally remastered versions of all the Beatles studio albums were released then.

For the Beatles search volume, the estimated first-order autocorrelation is 0.74 for the 2008–2009 period and 0.78 for the 2010–2011 period. For the Lady Gaga search volume, the estimated first-order autocorrelation is 0.98 for the 2008–2009 period and 0.73 for the 2010–2011 period. We could also estimate higher-order autocorrelations.

These are only estimates, but it would appear that the covariance stationarity assumption is badly violated for the Lady Gaga search volume time series. This assumption seems to be much more realistic for the Beatles search volume time series, although when we arbitrarily split the 2008–2011 data into two equal periods, there was greater variance in the first due to a single event that produced a large spike in searches.

When we talk below about trending, periodicity, and structural breaks, we will find that the assumption of covariance stationarity, necessary to estimate models of time series data, only needs to hold once we have controlled for the independent variables in the model—it is covariance stationarity conditioning on x_t that is relevant. Put another way, if the data-generating process is a covariance-stationary time series with a spike in search volume in the first half of September, our assumption of covariance stationarity is met as long as we control for that spike in our model. Failing to account for the spike may lead to a violation of covariance stationarity (Hendry, 2003, pp. 99–100).

It is important to note that if controls are necessary to meet the condition of covariance stationarity, we are restricted in our interpretation of the estimated model results for those controls. A term controlling for a nonstationary element should not be given any substantive interpretation beyond that of a control. Intervention analysis does soften this rule substantially—more on this in Chapters 3 and 5. If there are independent variables in our model to which we do wish to give substantive meaning and we do not think they are necessary to control for nonstationary elements, we would first need to confirm that the assumption of covariance stationarity is met without including these variables.

This is important, as it is not at all unusual to find that social science data, without any controls, violate the covariance stationarity assumption (Pickup, 2009). We have already seen this with the data used in the public responsiveness example in Chapter 1 (Figure 1.1), where the violation is due to trending—more on this to come. Any analysis of such data would need to take this problem into account.

2.5 Exogeneity

The usual exogeneity assumption in a regression model of cross-sectional data is the zero conditional mean assumption. This is described as follows:

$$y_i = \beta_0 + \beta_1 x_i + \varepsilon_i,$$

$$E(\varepsilon_i \mid X) = 0. \qquad (2.5.1)$$

This implies that the expected value of the error term is not a function of the explanatory variable(s) (Greene, 2002, p. 14). For models of time series data, it is also necessary to make an exogeneity assumption of this sort:

$$y_t = \beta_0 + \beta_1 x_t + \varepsilon_t,$$

$$E(\varepsilon_t \mid X) = 0. \qquad (2.5.2)$$

The bold X implies that the expected value of the error term in any given period is not a function of the explanatory variables in any time period— that is, at any lead or lag. For a single independent variable x_t, this can be written as

$$E(\varepsilon_t \mid x_{t+h}) = 0, \; \forall h. \qquad (2.5.3)$$

(*Note:* $\forall$ = "for all.") If this condition is met, x_t is said to be strictly exogenous for the estimation of β_1. What does it mean for the expected value of ε_t to be independent from an explanatory variable in the same or another time period? An obvious violation of independence is if ε_t and x_t are correlated (have covariance): $\text{COV}(\varepsilon_t, x_t) \neq 0$. The contemporaneous covariance between ε_t and x_t is

$$\text{Cov}(\varepsilon_t, x_t) = E((\varepsilon_t - \mu_\varepsilon)(x_t - \mu_X)). \qquad (2.5.4)$$

The covariance between ε_t and X_t from another time period is

$$\text{Cov}(\varepsilon_t, x_s) = E((\varepsilon_t - \mu_\varepsilon)(x_s - \mu_X)); \; s = t + h, \; h \neq 0. \qquad (2.5.5)$$

For example, $h = -1$:

$$\text{Cov}(\varepsilon_t, x_{t-1}) = E((\varepsilon_t - \mu_\varepsilon)(x_{t-1} - \mu_X)).$$

The presence of such covariance would violate the assumption of strict exogeneity. The result is a biased model estimate. A weaker assumption than strict exogeneity is that x_t is contemporaneously exogenous:

$$E(\varepsilon_t \mid x_t) = 0. \tag{2.5.6}$$

The expected value of ε_t is independent of x_t but is permitted to have some sort of dependence on lags or leads of x_t. Contemporaneous exogeneity will be sufficient for asymptotic unbiasedness in large T and requires the additional assumption of *weak dependence* (Wooldridge, 2006, pp. 382–384).[2] This will be discussed after we review some common violations of the strict-exogeneity assumption.

The violation of exogeneity is called endogeneity. This is a complex concept, but it is vital to understand it if we are interested in testing causal theories. But what could cause a violation of the zero conditional mean assumption? Let us say that we are interested in testing the causal effect of X_t on Y_t. Our data model is as follows:

For this discussion, we will limit ourselves to time series data. For other types of data, such as those described in Chapter 1, there are other possible data-generating processes and other endogeneity concerns. The following two data-generating processes will result in two forms of endogeneity. Each will complicate a test of the causal effect of X_t on Y_t.

1. An omitted variable causally prior to X_t and Y_t

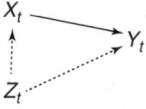

2. A simultaneous reciprocal relationship

A simultaneous reciprocal relationship is very difficult to deal with. It suggests that X_t and Y_t concurrently have causal effects on each other, not because of their relationship with any other variable but simply because they each cause each other at the same point in time. This is a very tricky problem and must be approached with advanced, multivariate time series techniques, such as vector autoregression. This is a topic for an advanced text on multivariate time series analysis (see Brandt & Williams, 2007). For our purposes, we will assume that the *contemporaneous* direction of causality runs in one direction and we know that direction. This leaves us with the omitted-variable problem.

Before discussing the different forms of the omitted-variable problem, let us examine it in general. Suppose the data-generating process is given as

$$y_t = \beta_0 + \beta_1 x_t + \beta_2 z_t + \varepsilon_t,$$

$$E(\varepsilon_t \mid X) = 0. \tag{2.5.7}$$

but we estimate the data model as

$$y_t = \tilde{\beta}_0 + \tilde{\beta}_1 x_t + \mu_t \tag{2.5.8}$$

and therefore, $\mu_t = \beta_2 z_t + \varepsilon_t$. The ordinary least squares (OLS) estimator of $\tilde{\beta}_1$ is

$$\tilde{\beta}_1 = \frac{\sum_{t=1}^{T} (x_t - \bar{x}) y_t}{\sum_{t=1}^{T} (x_t - \bar{x})^2}. \tag{2.5.9}$$

A demonstration that $\tilde{\beta}_1$ is an unbiased estimator of β_1, would be that $E(\tilde{\beta}_1 \mid X) = \beta_1$, where X denotes $\{x_1, x_2, \ldots, x_T\}$. In other words, the estimator is, on average, the true value.

If we take the expected value of Equation 2.5.9, conditioning on X, we can derive the following (Wooldridge, 2006):

$$E(\tilde{\beta}_1 \mid X) = \beta_1 + \left(\frac{1}{\sum_{t=1}^{T} (x_t - \bar{x})^2} \right) \sum_{t=1}^{T} (x_t - \bar{x}) E(\mu_t \mid X). \tag{2.5.10}$$

From $\mu_t = \beta_2 z_t + \varepsilon_t$, we can work out the following:

$$E(\mu_t \mid X) = E(\beta_2 z_t + \varepsilon_t \mid X),$$

$$= \beta_2 z_t. \tag{2.5.11}$$

Substituting Equation 2.5.11 into Equation 2.5.10,

$$E\left(\tilde{\beta}_1 \mid X\right) = \beta_1 + \left(\dfrac{1}{\sum_{t=1}^{T}\left(x_t - \overline{x}\right)^2}\right) \sum_{t=1}^{T}\left(x_t - \overline{x}\right)\beta_2 z_t,$$

$$= \beta_1 + \beta_2 \dfrac{\sum_{t=1}^{T}\left(x_t - \overline{x}\right)z_t}{\sum_{t=1}^{T}\left(x_t - \overline{x}\right)^2}. \qquad (2.5.12)$$

This reveals that $\tilde{\beta}_1$ may be a biased estimator of β_1, and the magnitude of the bias is

$$\beta_2 \dfrac{\sum_{t=1}^{T}\left(x_t - \overline{x}\right)z_t}{\sum_{t=1}^{T}\left(x_t - \overline{x}\right)^2}. \qquad (2.5.13)$$

The $\frac{\sum_{t=1}^{T}(x_t - \overline{x})z_t}{\sum_{t=1}^{T}(x_t - \overline{x})^2}$ term is the OLS estimator of the slope coefficient from the regression of z_t on x_t. This means that even if $\beta_1 = 0$, $E(\tilde{\beta}_1 \mid X)$ will not be so, except in two instances. If x_t and z_t are uncorrelated in the sample, then $\frac{\sum_{t=1}^{T}(x_t - \overline{x})z_t}{\sum_{t=1}^{T}(x_t - \overline{x})^2} = 0$, and the magnitude of the bias is 0. Alternatively, if $\beta_2 = 0$, the magnitude of the bias is 0. This amounts to saying that z_t is not actually in the data-generating process and so is not omitted from the data model.[3]

---

[3] Without going into the mathematics, omitting a variable from the data model may also have consequences for the estimated variance of the OLS estimator of β_1. If we use the usual estimate of the variance of an OLS estimator and compare the variance of $\tilde{\beta}_1$ from our misspecified data model (Equation 2.5.8) with the variance of $\tilde{\beta}_1$ from the correctly specified data model,

$$y = \hat{\beta}_0 + \hat{\beta}_1 x_1 + \hat{\beta}_2 x_2 + \varepsilon,$$

we will find that the estimated variance of $\tilde{\beta}_1$ (conditioning on x_1 and x_2) is less than that of $\tilde{\beta}_1$. This is true unless X_1 and X_2 are uncorrelated, in which case the variances are the same. The estimated variance of $\tilde{\beta}_1$ is too small. The conditioning on x_1 and x_2 means that this result may not hold in practice because of the potentially reduced estimate of the error variance due to the inclusion of x_2. As the sample size grows, the variance of each estimator shrinks to 0, making the difference in variance less important. However, remember that the misspecified model is biased unless $\beta_2 = 0$ or $\mathrm{Corr}(X_1, X_2) = 0$.

These consequences of omitting variables from the data model are easily extended to models with additional independent variables. The direction of bias is less clear, but the lesson is the same. If an independent variable in the data-generating process is omitted from the data variable and that omitted variable is correlated with an independent variable in the data model, the OLS estimator for the slope coefficient on the variable in the data model will be biased.

An omitted variable Z_t can be of different types, resulting from different data-generating processes:

1. Z_t is a third variable that is causally prior to and correlated with current values of X_t and Y_t.

2. Z_t is a past value of a third variable (Z_{t-1}) that is causally prior to and correlated with current values of X_t and Y_t.

3. Z_t is a past value of Y_t that is causally prior to and correlated with current values of X_t and Y_t.

The form of endogeneity from the first data-generating process is familiar to those who have studied the analysis of cross-sectional data. The forms of endogeneity due to the second and third data-generating processes are more specific to time series analysis, although those who have studied the use of instrumental variables in cross-sectional data may be familiar with these endogeneity problems (Greene, 2002, chap. 5). Let us use an example to discuss these endogeneity problems in a more concrete manner.

Let us say that Y_t is an aggregate measure of government vote intention, like that for the German government in Figure 2.1. Let us say that X_t is an aggregate measure of subjective evaluations of how the economy has performed. Our data model is the following:

We can now review the different types of data-generating processes that will lead to an omitted-variable bias:

1. Z_t is a third variable that is causally prior to and correlated with current values of X_t and Y_t,

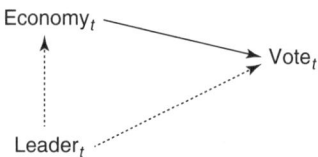

For example, both current government vote intention and current economic evaluations are a function of current leadership evaluations. Leadership effects on vote intention are well documented in a number of countries (Bittner, 2011; Johnston, 2002; Stewart & Clarke, 1992). Leadership effects on economic evaluations, such as those of prime ministerial and presidential approval, have also been demonstrated (Evans & Andersen, 2006; Evans & Pickup, 2010; Pickup & Evans, 2013). Concurrently, these will result in an apparent causal relationship between current economic evaluations and current government vote intention.

2. Z_t is a past value of a third variable (Z_{t-1}) that is causally prior to and correlated with current values of X_t and Y_t.

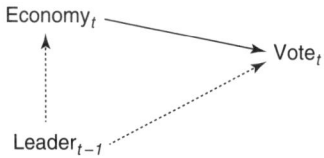

For example, both current government vote intention and current economic evaluations are a function of past leadership evaluations, resulting in an apparent relationship between current economic evaluations and current government approval. Work demonstrating the contemporaneous effects of leadership effects on economic valuations and vote intention is somewhat ambiguous as to whether it is contemporaneous leadership evaluations or earlier leadership evaluations that are relevant (Evans & Andersen, 2006).

3. Z_t is a past value of Y_t that is causally prior to and correlated with current values of X_t and Y_t.

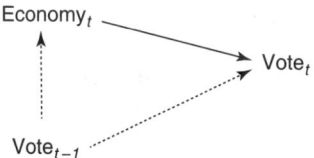

For example, both current government vote intention and current economic evaluations are a function of past government vote intention, resulting in an apparent causal relationship between current economic evaluations and current government approval. Recent work on economic evaluations makes the case that past vote intention influences current economic evaluations (Evans & Pickup, 2010; Ladner & Wlezien, 2007).

In practice, we can resolve these endogeneity problems by including the omitted variable. Of course, if the omitted variable is the past value of some variable in the context of cross-sectional analysis, there is not much we can do about it unless we have an appropriate instrumental variable to be used in an instrumental variable regression (see Angrist & Pischke, 2009). This demonstrates a key benefit of the analysis of longitudinal data, such as in time series analysis. We are much more likely to be in a position to include variables that would result in bias if omitted. Longitudinal data can be a powerful tool for addressing such endogeneity problems, which can go a long way in helping us test causal hypotheses.

It should be noted that the second endogeneity problem is a violation of strict exogeneity (Equation 2.5.2) but not contemporaneous exogeneity (Equation 2.5.6). It should also be noted that the third type of data-generating process will always lead to a violation of strict exogeneity, even if we include the omitted variable—the past value of Y_t. To see this second point, consider the following.

Our data-generating process is as follows:

$$\text{Vote}_t = \alpha_0 + \beta_1 \text{Econ}_t + \beta_2 \text{Vote}_{t-1} + \varepsilon_t,$$

$$\varepsilon_t \sim \text{NID}(0, \sigma_\varepsilon^2).$$

Our data model, estimated by OLS, is the following linear model (i.e., Vote_{t-1} is omitted):

$$\text{Vote}_t = \widetilde{\alpha}_0 + \widetilde{\beta}_1 \text{Econ}_t + \widetilde{\mu}_t.$$

Note: $\widetilde{\mu}_t$ denotes the OLS-estimated residuals from the model that excludes Vote_{t-1}; and $\widetilde{\beta}$ denotes the OLS-estimated coefficient that results from this exclusion. To see that this violates contemporaneous exogeneity, and $\widetilde{\beta}$ may be a biased estimate of β_1, we can rewrite the data-generating process as

$$\text{Vote}_t = \alpha_0 + \beta_1 \text{Econ}_t + \mu_t,$$

$$\mu_t = \beta_2 \text{Vote}_{t-1} + \varepsilon_t.$$

The data-generating process for μ_t is a function of Vote_{t-1}. Therefore, if Econ_t is also a function of Vote_{t-1} in the data-generating process, then

$$E(\mu_t \mid \text{Econ}_t) = E(\beta_2 \text{Vote}_{t-1} + \varepsilon_t \mid \text{Econ}_t) = \beta_2 \widetilde{\delta}_1,$$

where $\widetilde{\delta}_1$ is the slope coefficient from the regression of Vote_{t-1} on Econ_t.

Therefore, we do not have contemporaneous exogeneity: $E(\mu_t | \text{Econ}_t) \neq 0$; and our estimate of β_1 will be biased. This may be resolved by explicitly modeling Vote_{t-1} as in the following data model, again estimated by OLS:

$$\text{Vote}_t = \hat{\alpha}_0 + \hat{\beta}_1 \text{Econ}_t + \hat{\beta}_2 \text{Vote}_{t-1} + \hat{\varepsilon}_t.$$

The residual will no longer contain Vote_{t-1}, so we have contemporaneous exogeneity:

$$E(\varepsilon_t | \text{Econ}_t) = 0.$$

However, we will not have strict exogeneity. This is because Vote_t is, by the definition of the data-generating process, a function of ε_t, and vice versa. Therefore, ε_t is a function of a lead or lag of the independent variable Vote_{t-1}. Specifically, ε_t is a function of the first lead of Vote_{t-1}:

$$E(\varepsilon_t | \text{Vote}_{t-1+1}) = E(\varepsilon_t | \text{Vote}_t) \neq 0.$$

This is a problem, but in Chapter 3, we will see that if an assumption called weak dependence can be met and we have contemporaneous exogeneity, even if we do not have strict exogeneity, we can get asymptotically unbiased estimates of the parameters in our model. Before discussing weak dependence, let us make a brief foray into a more complicated discussion of exogeneity.

We have been defining exogeneity in terms of expectations and covariances, and will continue to do so. However, Engle, Hendry, and Richard (1983) have shown that this can lead to an ambiguity. Because of the ambiguity that can occur when defining exogeneity in terms of expectations, Engle et al. (1983) define what they call weak exogeneity as the necessary exogeneity condition for an unbiased estimation of the parameters in a model. They also define two more restrictive conditions called strong exogeneity and super exogeneity, for the purposes of forecasting and intervention analysis, respectively. Weak exogeneity resolves the ambiguity problem but very much increases the necessary complexity of the discussion. Even the relatively advanced econometrics text by Greene (2002) provides a brief definition before referring the interested reader to what is described as challenging reading (p. 381).[4] An incomplete description is provided here.

For our purposes, we consider the model of y_t as a function of x_t. We begin by denoting the joint density for y_t and x_t as $f(y_t, x_t)$; and the

---

[4] In addition to Engle et al. (1983), these advance readings include Zellner (1979), Sims (1977), and Granger (1969).

distribution of y_t conditional on x_t and a set of parameters $\boldsymbol{\beta}$ as $f(y_t \mid \boldsymbol{\beta} x_t)$. The conditional distribution $f(y_t \mid \boldsymbol{\beta} x_t)$ is our model. Also, we denote the marginal distribution of x_t as $f(x_t \mid \boldsymbol{\theta})$, where $\boldsymbol{\theta}$ is the set of parameters of the marginal distribution. The marginal distribution $f(x_t \mid \boldsymbol{\theta})$ is the data-generating process for x_t. Weak exogeneity of x_t for the estimation of $\boldsymbol{\beta}$ in our model is established if we can write the joint distribution for y_t and x_t as

$$f(y_t, x_t) = f(y_t \mid \boldsymbol{\beta} x_t) f(x_t \mid \boldsymbol{\theta}) \qquad (2.5.14)$$

such that no element of $\boldsymbol{\beta}$ is functionally related to $f(x_t \mid \boldsymbol{\theta})$. This means that the parameters of $f(x_t \mid \boldsymbol{\theta})$ can be safely ignored in the estimation of the parameters of $f(y_t \mid \boldsymbol{\beta} x_t)$. Note that the exogeneity of x_t is established with respect to a specific set of parameters of interest in a specific model.

2.6 Weak Dependence

A stationary time series is weakly dependent if Y_t and Y_{t+h} are "almost independent" as h increases. If for a covariance stationary process $\mathrm{Corr}(y_t, y_{t+h}) \to 0$ as $h \to \infty$, we say that this covariance stationary process is weakly dependent.

Say we have a time series with the following data-generating process:

$$y_t = \alpha_1 y_{t-1} + \varepsilon_t, \text{ with } \varepsilon_t \sim \mathrm{NID}\left(0, \sigma_\varepsilon^2\right). \qquad (2.6.1)$$

This is the autoregressive process of order 1 (AR(1)) described above. For this process to be weakly dependent, it must be the case that $|\alpha_1| < 1$. This is because for an AR(1) process, $\mathrm{Corr}(y_t, y_{t+h}) = \alpha_1^h$, which becomes small as h increases if $|\alpha_1| < 1$.[5]

This suggests one example of a time series process that is not weakly dependent. In Equation 2.6.1, set α_1 is equal to 1:

$$y_t = y_{t-1} + \varepsilon_t. \qquad (2.6.2)$$

For such a process, α_1 is no longer interpretable as the correlation between itself and the first lag of itself, and $\mathrm{Corr}(y_t, y_{t+h}) \nrightarrow 0$ as $h \to \infty$. This particular process is called a unit root; we will revisit it at the end of this chapter and in Chapter 6.

---

[5] In fact, Equation 2.6.1 is not an autoregressive process unless $|\alpha_1| < 1$.

When it is met, the assumption of weak dependence, combined with the assumption of stationarity, allows us to weaken the strict-exogeneity assumption to contemporaneous exogeneity for large-T analysis. Under these conditions, the sample variances and covariances converge on the population variances and covariances as T goes to infinity. In this case the β_1 from the regression of y_t on x_t,

$$y_t = \beta_0 + \beta_1 x_t + \varepsilon_t,$$

can be estimated by

$$\hat{\beta}_1 = \frac{\widehat{Cov(Y_t, X_t)}}{\text{Var}(X_t)}. \tag{2.6.3}$$

This estimate is asymptotically unbiased and consistent as $T \to \infty$.

2.7 Trending, Periodicity, and Structural Breaks

We saw in Chapter 1 that a trending time series is one in which the mean increases or decreases over time. Recall that for a time series to be covariance stationary, its mean must be constant over time, as do its variance and (auto)covariances. Therefore, a trending time series cannot be stationary, since the mean is changing over time.

As in the public responsiveness example from Chapter 1, social science time series often have trends, and it is important to spot this form of violation of the covariance stationarity assumption. If two series are trending together, we will probably estimate a strong correlation between the two, but it is highly likely that it is a spurious result produced by the fact that both variables are a function of time. Often, both will be trending because of other, unobserved, factors.

The easiest way to spot a trending time series, in the first instance, is to plot and examine the data. When considering whether a time series trends or not, it is important to keep in mind that there are actually a number of dynamic processes that are called trending. The fact that a time series is trending is often not difficult to spot. It is a little harder to determine the type of trend. Let us consider the simplest, a deterministic, linear trend. (We will discuss other types of trends and how to distinguish between them in Chapter 6.) An example of a data-generating process with a deterministic, linear trend is the following:

$$y_t = \beta_0 + \beta_2 t + \varepsilon_t \tag{2.7.1}$$

In this equation, t is a variable that counts up from the beginning of the time series: $t = \{1, 2, 3, \dots, T\}$. The interpretation of such a process is that y_t

increases or decreases, on average, by the magnitude of β_2 in each time period.[6] The average growth or decline of y_t may not remain constant in the long run, but if it does so for the period of observation, then this is an acceptable way to express the data-generating process. Otherwise, a more complex function of time may be used to express a nonlinear trend. A quadratic trend is one option. Splines are a more advanced option (Durlauf & Blume, 2010; Keele, 2008).

A deterministic linear trending series can be weakly dependent—the correlation between y_t and y_{t+h} approaches 0 as h increases. If a series is weakly dependent and is stationary about its trend, we call it a trend-stationary process. When we state that the series is stationary about its trend, we mean that the series is stationary once we have partialled out the trend. If the time series is trend stationary, we can control for the trend to avoid a spurious result. As stated previously, the assumption of covariance stationarity, necessary to estimate models of time series data, only needs to hold once we have controlled for the independent variables in the model. It is covariance stationarity conditioning on X_t that is the relevant assumption. Let us return to the public responsiveness example from Chapter 1.

$$R_t = \beta_0 + \beta_1 P_t + \beta_2 W_t + \varepsilon_t. \qquad (2.7.2)$$

Recall that R_t is the public's relative preference for policy spending in a given year—that is, the difference between the public's preferred level of policy spending and the level that it actually gets. P_t is the actual level of policy spending in a year. W_t represents other, exogenous effects on the public's relative preferences. This is an example of a static model, which we shall cover in Chapter 3.

In the Canadian model for social welfare spending developed by Soroka and Wlezien (2010), W_t is a counter variable: $\{1, 2, 3, \ldots\}$. It is included to capture the upward trend in spending demands. The counter variable is just a trend variable, t. The results of regressing the public's relative spending preference on the level of government spending and the counter variable are presented in Table 2.1.

The results indicate a statistically significant negative relationship between government spending and the public's relative spending preference. For each billion-dollar increase in government spending, the public's relative preference for spending decreases by 0.61 percentage points. As government spending goes up, the public's preference for greater spending

---

[6] Deviations from this average increase/decrease are modeled as random.

Table 2.1   Canadian Public Responsiveness Model, Social Welfare
Spending

Preference	Coefficient	Standard Error	t Statistic	P Value
Program spending	−0.61	0.14	−4.46	0.001
Counter	3.95	0.50	7.89	<0.001
Constant	117.51	22.27	5.28	<0.001

NOTE: $R^2 = 0.87$, $T = 16$; T = number of time points.

decreases. This is the result predicted by the thermostatic model. What if Soroka and Wlezien had not included the trend variable? We already noted in Chapter 1 that both the public's relative preference for social welfare spending and the actual social welfare spending are trending upward over time. This means that there is a danger of estimating a positive relationship between the two variables simply because they both trend upward and not because there is a positive causal relationship.

This is indeed what we found when we estimated such a model without a trend in Chapter 1. We found a positive relationship between government social welfare spending and the public's relative preference. Controlling for the trending in the data has clear consequences for our analysis and the conclusions we reach about the relationship between government spending and the public's relative preference for spending. If we were choosing between these two models, we would choose the model with the trend. We can see from the statistical significance of the trend term in Table 2.1 that the public's relative preference does contain a trend. The estimates from the model that fails to control for this trend violate the assumptions of covariance stationarity. As we will discuss in Chapter 6, a deterministic, linear trend can also be removed by first differencing the data. However, this has consequences that generally make it an unattractive option.

Less common than trending but still common is the problem of data containing periodicity. This is a time series process with an equilibrium that cycles up and down. If it cycles up and down in time with the seasons, it is often called seasonality. Examples include quarterly data on retail sales, which tend to jump up in the fourth quarter, and crime rates, which tend to be higher in warmer months. Seasonality can be dealt with by adding a set of seasonal or monthly dummies. We can also seasonally difference the data. For example, if we have monthly data and we wish to control for periodicity at the monthly level, 12th differencing the data (subtracting the

12th lag of each variable from itself) might eliminate the periodicity resulting in stationary data. Substantively speaking, this transforms the data into the change since the same month in the previous year. This is discussed further in Chapter 6.

Trending and seasonality are instances where the mean of the time series is changing with time and is therefore not stationary. The mean could also change in a less systematic way. It could simply shift to a new level at a particular time. This shift is called a structural break or equilibrium shift. If this equilibrium shift is explained by an exogenous variable in our model, then all is well. Otherwise, an unaccounted-for equilibrium shift will result in biased estimations of our data model parameters.

In the German government approval/vote intention example in Chapter 1, we saw an example of a structural break in the variance of the time series. As noted, structural breaks can also occur in the covariance between time series. This means that the relationship between two time series variables might change at a specific point in time. Such a possibility is both something that needs to be taken into account and possibly of substantive interest. We shall discuss this further in the context of intervention analysis in Chapter 5. In Chapter 3, we will discuss other ways in which we can test and control for a potential deterministic trend, seasonality, or structural breaks.

2.8 Instability and Integration

There are reasons other than trending, seasonality, and structural breaks that will cause a time series to be nonstationary. The most important of these is instability. Stability relates to the concepts of convergence and equilibrium. A time series is stable if it has an equilibrium to which it converges in the long run.

Consider the, now familiar, AR(1) process:

$$y_t = \alpha_0 + \alpha_1 y_{t-1} + \varepsilon_t. \tag{2.8.1}$$

We noted that this process is weakly dependent if $|\alpha_1| < 1$. This is also a requirement for this process to be stable. As stability is a necessary condition for stationarity, the AR(1) process cannot be stationary unless $|\alpha_1| < 1$.

To see why $|\alpha_1| < 1$ is a necessary condition for stability, let us generate one realization of the stochastic $\{\varepsilon_t\}$ sequence and plot one realization of the data-generating process, described by Equation 2.8.1, for different values of α_1. For simplicity, let $\alpha_1 = 0$ and $\varepsilon_t \sim N(0, 1)$. Now let us consider the cases where α_1 takes on the values 0.9, 0.5, −0.5, 1.2, and −1.2. We are interested in determining in which of these cases the time series process is stable and therefore meets a necessary condition for stationarity. We do this by examining Figures 2.4a to f. In each plot, Series A is our randomly generated time series. Series B is the nonstochastic part of A: $y_t = \alpha_1 y_{t-1}$. Series A moves stochastically around the nonstochastic part—that is, Series B. The nonstochastic series gives us an impression of the value to which the time series will converge (or not) in the long run ($t \to \infty$), when the stochastic component becomes arbitrarily small ($\varepsilon_t \to 0$).

Figure 2.4a is the plot of $y_t = \alpha_0 + \alpha_1 y_{t-1} + \varepsilon_t$, where $\alpha_0 = 0$ and $\alpha_1 = 0.9$. We have also set the initial value for y_t at $y_{t=1} = 0$. In this plot, you can see from the nonstochastic part of the series that the time series process converges to (remains in) an equilibrium of 0. The equilibrium value also happens to be the initial value, so the series is not so much converging on its equilibrium as it is remaining at its equilibrium. An example of this type of convergent sequence might be monthly approval ratings for the U.S. president (although α_0 would probably not equal 0 and all values would be positive).

We can get a clearer picture of convergence by choosing an initial value that is out of equilibrium, for example, $y_{t=1} = 1$. This is plotted in Figure 2.4b. The monthly approval of the U.S. president could also possibly be an example of this type of sequence. Presidents often start their term with a honeymoon period in which their support is atypically high—that is, a president's approval often starts out of equilibrium.

The comparison of Figures 2.4a and b raises an interesting point. The time series processes plotted in these two figures are the same, except that we began the second one out of equilibrium. Both of these processes are stable in that they converge to an equilibrium in the long run. However, the series that started out of equilibrium is not stationary. As is clear from the plot, it trends downward until it reaches its equilibrium. Having done so, the process continues as a stationary process.[7] If we observed this process as it equilibrated, it would not be stationary. If we observed this process at a later point in time, long after its initial out-of-equilibrium start, it would be stationary. We will discuss the necessary conditions for stationarity further in Chapters 5 and 6. For now it is important to note that stability is a necessary but not sufficient condition for stationarity.

Now consider Figures 2.4c and d, in which α_1 is 0.5 and −0.5, respectively. Again, we start the time series processes out of equilibrium: $y_{t=1} = 1$. Both of these series are stable in that they converge to their equilibrium in the

---

[7] Such a process might be called asymptotically stationary.

Figure 2.4   Stable and Unstable Time Series Processes

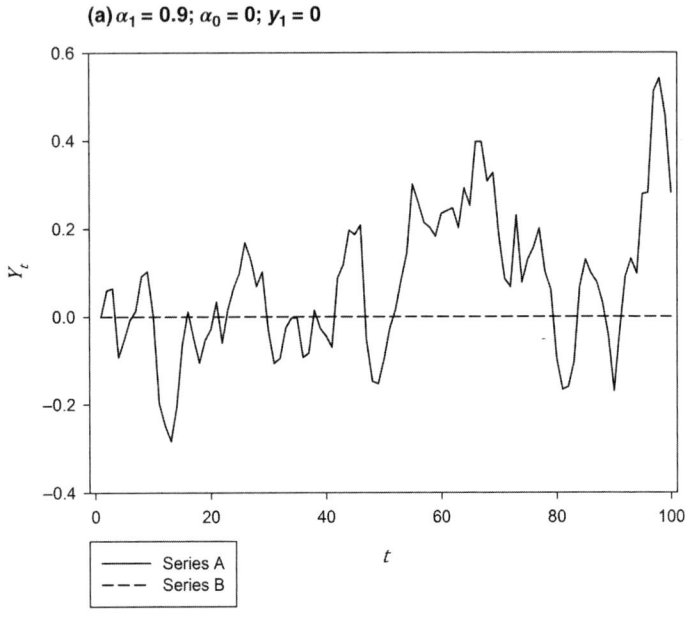

(a) $\alpha_1 = 0.9$; $\alpha_0 = 0$; $y_1 = 0$

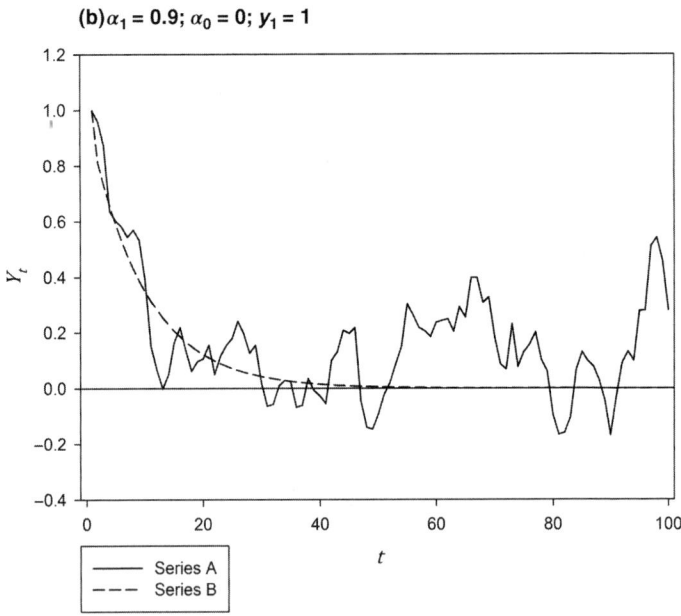

(b) $\alpha_1 = 0.9$; $\alpha_0 = 0$; $y_1 = 1$

Figure 2.4

(c) $\alpha_1 = 0.5$; $\alpha_0 = 0$; $y_1 = 1$

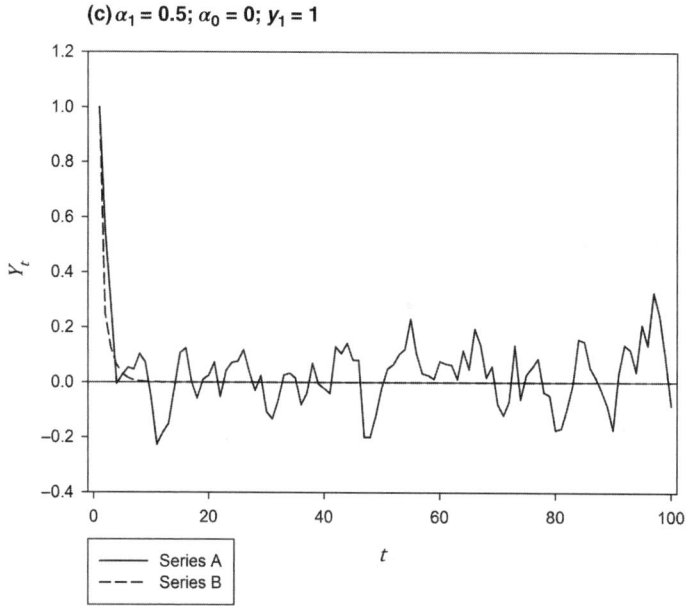

(d) $\alpha_1 = -0.5$; $\alpha_0 = 0$; $y_1 = 1$

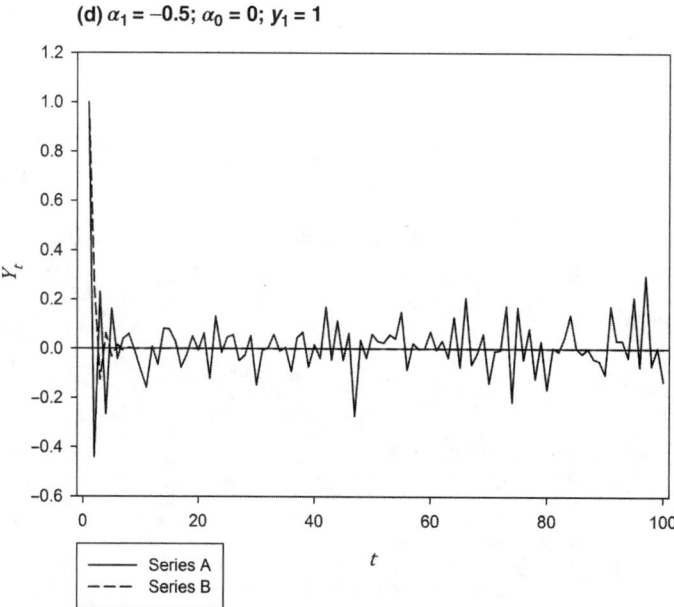

Figure 2.4 (Continued)

(e) $\alpha_1 = 1.2$; $\alpha_0 = 0$; $y_1 = 1$

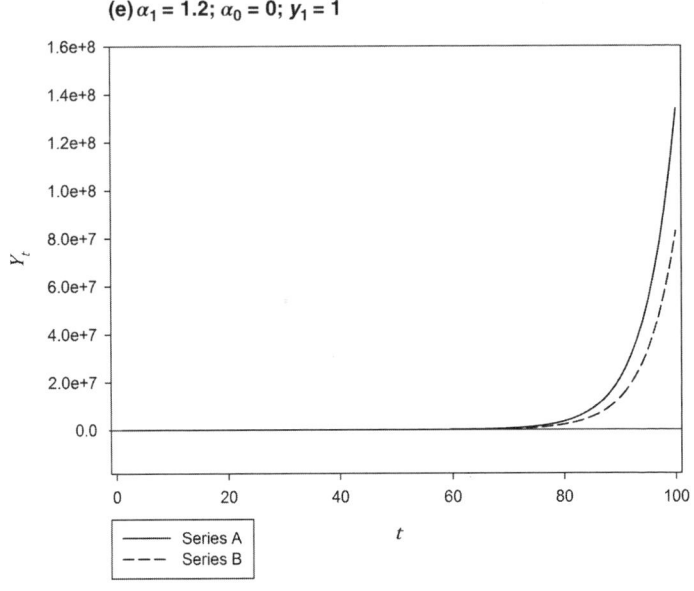

(f) $\alpha_1 = -1.2$; $\alpha_0 = 0$; $y_1 = 1$

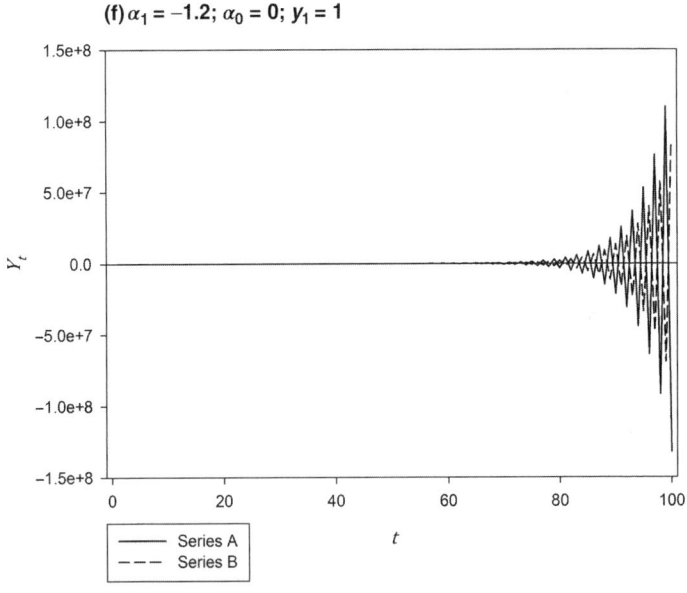

long run, but the second series is a stable, oscillating process. It converges on its equilibrium through a decaying, oscillating path.

Each of the series with $|\alpha_1| < 1$ converged to an equilibrium. The equilibrium of such series can be calculated as

$$E\left(y_t\right) = \frac{\alpha_0}{\left(1-\alpha_1\right)} \cdot \tag{2.8.2}$$

For example, when $\alpha_0 = 0$ and $\alpha_1 = 0.9$:

$$E\left(y_t\right) = \frac{\alpha_0}{\left(1-\alpha_1\right)} = \frac{0}{\left(1-0.9\right)} = 0.$$

The convergence to equilibrium is different for each of the processes considered so far. The time series process with $\alpha_1 = 0.5$ converges quicker than for $\alpha_1 = 0.9$, and $\alpha_1 = -0.5$ oscillates around its equilibrium before settling down. An example of a convergent series with $0 < \alpha_1 < 1$ is changes in the yearly amount of alcohol consumed in Denmark (liters per capita): $y_t = \alpha_0 + \alpha_1 y_{t-1} + \varepsilon_t$, where y_t is the yearly change in the log of the amount of alcohol consumed, $\hat{\alpha}_1 = 0.806$, $SE = 0.15$. (Bentzen & Smith, 2004).[8]

Convergent series with $\alpha_1 < 0$ are far less common, particularly in the social sciences. However, everyone is familiar with a pendulum (Figure 2.5). The horizontal distance that the bob of the pendulum travels is a classic example of a stable system that oscillates into its equilibrium after it has been started out of equilibrium. If disturbed, the pendulum will swing left and right until gravity returns it to its original position. Gravity moves the pendulum back to equilibrium through an oscillation.

Convergent sequences reflect stable systems. Let us now look at systems that are not stable.

Specifically, let us consider the same process: $y_t = \alpha_0 + \alpha_1 y_{t-1} + \varepsilon_t$, where $y_0 = 1$ and $\alpha_0 = 0$ but $|\alpha_1| > 1$. Figures 2.4e and f correspond to $\alpha_1 = 1.2$ and -1.2, respectively. The time series process in Figure 2.4e is an explosive process. The time series process in Figure 2.4f is an oscillatory explosive process. As these processes do not return to any equilibrium, they are not stable, and therefore they are not stationary.

When discussing weak dependence, we considered the possibility that $\alpha_1 = 1$. This produces a time series process with very special properties. As with the case where $|\alpha_1| > 1$, the time series does not converge to an equilibrium and is not stationary. However, the process is not explosive. This is an example of a time series process called a unit root process.

---

[8] The Bentzen and Smith model also includes the price of alcohol.

Figure 2.5 Pendulum

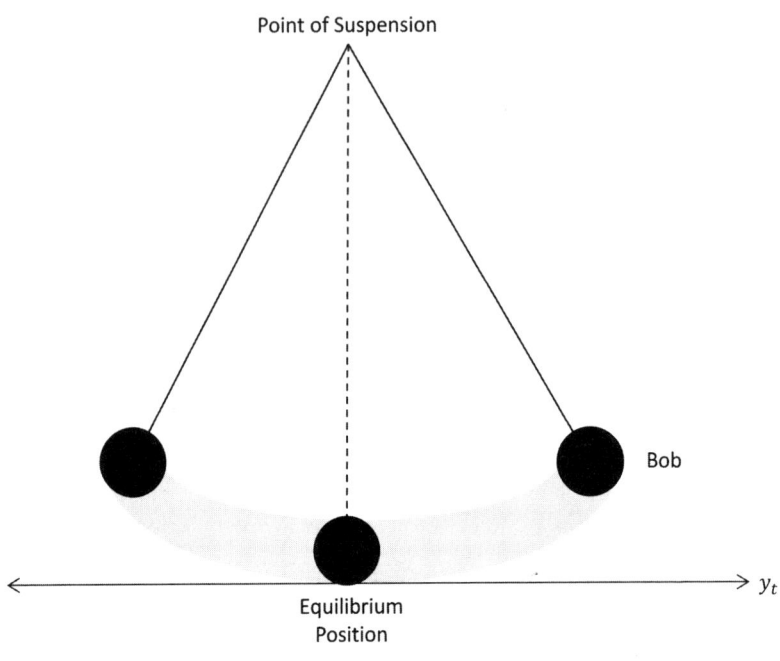

A unit root process is an example of a category of time series processes we call integrated. An integrated time series process is one that is not stationary but can be made stationary by differencing one or more times. A unit root process can be made stationary by differencing once as follows:

$$y_t = \alpha_0 + y_{t-1} + \varepsilon_t. \tag{2.8.3}$$

Subtract y_{t-1} (the lag of y_t) from both sides:

$$y_t - y_{t-1} = \alpha_0 + y_{t-1} - y_{t-1} + \varepsilon_t,$$

$$\Delta y_t = \alpha_0 + \varepsilon_t. \tag{2.8.4}$$

As $\varepsilon_t \sim \text{NID}(0,\sigma_\varepsilon^2)$, Δy_t is a constant plus a white noise term. This process is stable and meets the conditions of covariance stationarity. The substantive interpretation of this process is that it has infinite memory. Consider the case when $\alpha_0 = 0$. When something intervenes to change y_t, the process, in the absence of any further intervention, remains at its new value. In other words, the effect of the intervention remains forever. We will consider this and other integrated time series processes in detail in Chapter 6.

Summary

In this chapter, you have been introduced to fundamental concepts in time series data and analysis. These concepts are the basis of the analytical techniques discussed in the chapters to follow. In the next chapter, you will see how each of these fundamental time series concepts are applied as we examine how time series data can be modelled using two of the most basic time series models: static models and finite distributed lag models.

CHAPTER 3: STATIC TIME SERIES MODELS AND ORDINARY LEAST SQUARES ESTIMATION

Having covered many of the fundamental concepts of time series in Chapter 2, in this chapter we begin to explore basic models of time series data. This chapter examines the static and finite distributed lag (FDL) models estimated through ordinary least squares (OLS) regression. We examine the assumptions required for the estimation of such models to be unbiased. We also examine how to test for and correct violations of key assumptions—in particular covariance stationarity and no serial correlation. In the discussion of violations of covariance stationarity, we discuss the challenges of trending, periodicity, and structural breaks. In the discussion of correcting for serial correlation, we are introduced to time series models that are estimated by maximum likelihood.

3.1 Static and Finite Distributed Lag (FDL) Models

Let us consider two simple time series models. A *static* time series model relates contemporaneous variables. For example, with one independent variable,

$$y_t = \beta_0 + \beta_1 x_t + \varepsilon_t. \tag{3.1.1}$$

An FDL model allows one or more variables to affect y_t with a lag. For example, with one independent variable and two of its lags,

$$y_t = \beta_0 + \delta_0 x_t + \delta_1 x_{t-1} + \delta_2 x_{t-2} + \varepsilon_t. \tag{3.1.2}$$

In the above, δ_0, δ_1, and δ_2 are slope parameters just like β_1. More generally, an FDL model of order q will include q lags of x_t: $(x_{t-1}, x_{t-2}, \ldots, x_{t-q})$. We call δ_0 the impact propensity (effect)—it reflects the immediate change in y_t due to a one-unit change in x_t. For a temporary, one-period change, y_t returns to its original level q periods after x_t returns to its original value. To get a sense of what this means, consider the $q = 2$ FDL process:

$$y_t = \beta_0 + \delta_0 x_t + \delta_1 x_{t-1} + \delta_2 x_{t-2} + \varepsilon_t. \tag{3.1.3}$$

Let us assume that $x_t = 0$ at $t = 1$ and $t = 2$, then, $x_t = 1$ at $t = 3$, and x_t returns to 0 thereafter. From Equation 3.1.3, we can calculate the expected value of y_t, conditioning on x_t, x_{t-1}, x_{t-2} for $t = 1$ to $t = 7$. The calculated values are given in Table 3.1.

We can graphically display the same results, as in Figure 3.1.

Table 3.1 Expected Values From a $q = 2$ FDL Process

t	x_t	$E(y_t \mid x_t, x_{t-1}, x_{t-2})$
1	0	β_0
2	0	β_0
3	1	$\beta_0 + \delta_0$
4	0	$\beta_0 + \delta_1$
5	0	$\beta_0 + \delta_2$
6	0	β_0
7	0	β_0

NOTE: FDL = finite distributed lag.

Figure 3.1 Expected Values From a $q = 2$ FDL Process

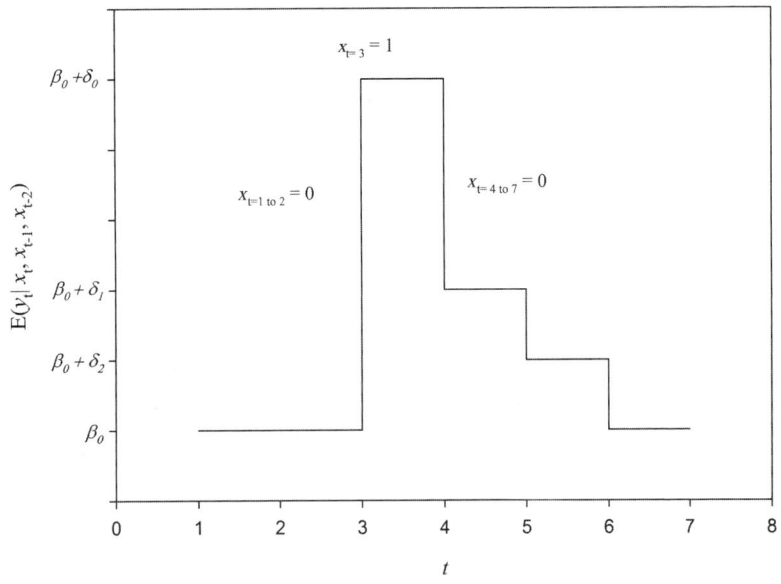

NOTE: FDL = finite distributed lag.

Both the table and the figure demonstrate the same thing. With $x_t = 0$ for the first two time points, the expected value of y_t is the constant β_0. When x_t increases by one unit to 1 at $t = 3$, the expected value increases to $\beta_0 + \delta_0$. The change, δ_0, is the coefficient on x_t and is the impact propensity. At $t = 4$, x_t returns to 0. However, x_{t-1} is the value of x_t at the previous time point, and so $x_{t-1} = 1$ at $t = 4$. This means that the expected value is now the constant plus the coefficient on x_{t-1}: $\beta_0 + \delta_1$. Note that the graphical representation assumes that $\delta_1 < \delta_0$. This was just chosen for the purpose of illustration, and nothing requires it to be the case. At $t = 5$, x_t remains 0, and x_{t-1} now returns to 0. Since x_{t-2} is the value of x_t two times previous, $x_{t-2} = 1$ at $t = 5$. The expected value is now the constant plus the coefficient on x_{t-2}: $\beta_0 + \delta_2$. At $t = 6$, x_t and x_{t-1} remain 0, and x_{t-2} returns to 0. The expected value is now simply back to the constant: β_0.

We call $\delta_0 + \delta_1 + \cdots + \delta_q$ the long-run propensity or long-run effect. It reflects the long-run change in y_t after a permanent one-unit change in x_t. Let us assume that $x_t = 0$ at $t = 1$ and $t = 2$; then, $x_t = 1$ thereafter. Again from Equation 3.1.3, we can calculate the expected value of y_t, conditioning on x_t, x_{t-1}, x_{t-2} for $t = 1$ to $t = 7$. The calculated values are given in Table 3.2 and graphically displayed in Figure 3.2.

Again, the table and figure demonstrate the same thing. With $x_t = 0$ for the first two time points, the expected value of y_t is the constant β_0. When x_t

Table 3.2 Expected Values From a $q = 2$ FDL Process

t	x_t	$E(y_t \mid x_t, x_{t-1}, x_{t-2})$
1	0	β_0
2	0	β_0
3	1	$\beta_0 + \delta_0$
4	1	$\beta_0 + \delta_0 + \delta_1$
5	1	$\beta_0 + \delta_0 + \delta_1 + \delta_2$
6	1	$\beta_0 + \delta_0 + \delta_1 + \delta_2$
7	1	$\beta_0 + \delta_0 + \delta_1 + \delta_2$

NOTE: FDL = finite distributed lag.

Figure 3.2   Expected Values From a $q = 2$ FDL Process

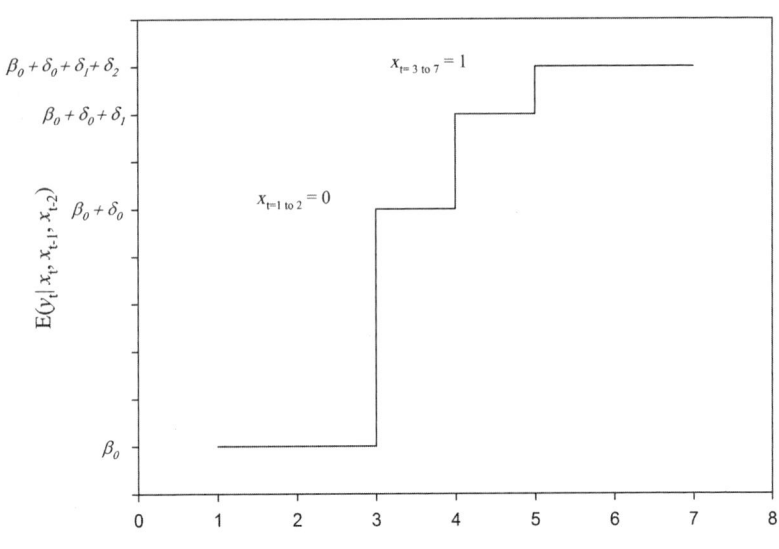

NOTE: FDL = finite distributed lag.

increases by one unit to 1 at $t = 3$, the expected value increases to $\beta_0 + \delta_0$. At $t = 4$, x_t remains 1 and x_{t-1}, the value of x_t at the previous time point, is also 1. This means that the expected value is now the constant plus the coefficient on x_t plus the coefficient on x_{t-1}: $\beta_0 + \delta_0 + \delta_1$. At $t = 5$, x_t and x_{t-1} remain 1, and x_{t-2}, the value of x_t two time points prior, is 1. The expected value is now the constant plus the coefficient on x_t plus the coefficient on x_{t-1} plus the coefficient on x_{t-2}: $\beta_0 + \delta_0 + \delta_1 + \delta_2$. At $t = 6$, x_t, x_{t-1}, and x_{t-2} remain 1, and the expected value remains as $\beta_0 + \delta_0 + \delta_1 + \delta_2$. As long as x_t does not change, the expected value will not change. The difference between this expected value and the expected value before the one-unit change in x_t at $t = 3$ is the full effect of this change: $\delta_0 + \delta_1 + \delta_2$. This is the long-run propensity.

We can compare the FDL process with the static process (no lags of the independent variables). In the static process (Equation 3.1.1), the estimated short-run and long-run effects are the same and equal: β_1. A permanent one-unit change in x_t produces an immediate effect of β_1 on y_t, and this is the sum total of its estimated effect.

Both the static process and the FDL process can contain multiple independent variables. For example, consider an FDL process that has k independent variables each with q lags:

$$y_t = \beta_0 + \beta_{1,0}x_{1,t} \cdots + \beta_{k,0}x_{k,t} + \sum_{i=1}^{k}\sum_{m=1}^{q}\beta_{i,m}x_{i,t-m} + \varepsilon_t \quad (3.1.4)$$

Note that if we do have more than one independent variable, these can each be included with a different number of lags. Both the static model and the FDL model can be estimated using OLS. Just as with the application of OLS to cross-sectional data, there are some assumptions that need to be met when applying OLS to time series data. It is to these assumptions that we now turn.

3.2 Assumptions for Unbiasedness With Time Series Analysis

Many of the assumptions necessary for OLS regression with cross-sectional data apply in the same way as when we are using time series data. We assume that the relationships between Y_t and the X_ts are linear. We assume that no X_t is constant and there is no perfect collinearity. We will also need to make a zero conditional mean assumption, as defined for time series data in Section 2.5 in Chapter 2. But what about the assumption of a random sample?

As discussed in Chapter 2, time series data are not produced by repeatedly sampling randomly from a population or a data-generating process. Time series data are a single realization of a random data-generating process. We do not have a random sample of realizations of Y_t with which we can estimate $E(y_t)$ for a particular time point t. We can, however, estimate $E(Y_t)$ based on the average value of a single realization of Y_t across all t. This is useful if we can assume that this converges on $E(Y_t)$ when you have enough time points:

$$\text{As } T \to \infty: E(Y_t) \to E(y_t) \text{ for all } t, t = 1, \ldots, T \quad (3.2.1)$$

As we saw in Chapter 2, this will be true if y_t is covariance stationary. Recall that a stochastic process is covariance stationary if $E(y_t)$ is constant, $\text{Var}(y_t)$ is constant, and, for all $k > 0$ and $h \neq 0$, $\text{Cov}(y_{t=k}, y_{t=k+h})$ depends only on h and not on k. This allows us to estimate the means, variances, and covariances of the data-generating process from the observed data.

In summary, covariance stationarity makes it possible to use the estimated parameters from the data model based on the single observed realization to say something about the unobserved data-generating process. This is equivalent to the idea in cross-sectional data analysis that having a random sample of data means that the sample is representative of the larger population, and so we are able to say something about the unobserved population from the parameters estimated from the observed data model.

Assumptions about the representativeness of the means, variances, and covariances estimated from the data are the same in both cases, although with cross-sectional data this is sometimes expressed in terms of the population. For time series data, it is always expressed relative to the data-generating process. For cross-sectional data, this representativeness is often derived from the assumption that the data are the product of random, independent draws from the population or data-generating process. For time series data, this representativeness is derived from the assumption of stationarity.

Just as in the cross-sectional case, we need to add an assumption of homoskedasticity in order to be able to derive the correct standard errors from the usual OLS estimator. For time series data, we assume that $\text{Var}(\varepsilon_t|x_s) = \sigma^2$, for all s. This means that the error variance is independent of the xs for all leads and lags and it is constant over time.

To derive the correct standard errors, when using time series data, we also need the assumption of no serial correlation, conditional on X. As noted in Chapter 2, serial correlation is another name for autocorrelated errors:

$$\text{Corr}\left(\varepsilon_t, \varepsilon_{t-h} \mid X\right) = \frac{\text{Cov}\left(\varepsilon_t, \varepsilon_{t-h} \mid X\right)}{\text{Var}\left(\varepsilon_t, \varepsilon_t \mid X\right)}. \tag{3.2.2}$$

The conditioning on X is often left implicit. For example, first-order serial correlation is the correlation of the error with the first lag of itself:

$$\text{Corr}\left(\varepsilon_t, \varepsilon_{t-1}\right) = \frac{\text{Cov}\left(\varepsilon_t, \varepsilon_{t-1}\right)}{\text{Var}\left(\varepsilon_t, \varepsilon_{t-1}\right)}. \tag{3.2.3}$$

The no serial correlation assumption is as follows:

$$\text{Corr}(\varepsilon_t, \varepsilon_{t+h}) = 0 \text{ for } h \neq 0. \tag{3.2.4}$$

In total, we have five assumptions:

A.1 Linear in parameters

A.2 Variance in all X and no perfect collinearity

A.3 Zero conditional mean (strict exogeneity)

A.4 Homoskedasticity

A.5 No serial correlation

Under these five assumptions and the assumption of covariance stationarity, the OLS estimator of the regression coefficients is unbiased, and the OLS estimator of σ^2 in the time series case is the same as in the cross-sectional case. These assumptions are called the Gauss-Markov assumptions. With the additional assumption of normal and independent errors, inference is the same as in the cross-sectional case.

A.6 Normal and independent errors

Together, these six assumptions are called the classical linear model assumptions of time series regression (Wooldridge, 2009). Let us think about this more concretely by returning to our example of public responsiveness in Canada from Chapter 1. Our model is as follows:

$$R_t = \beta_0 + \beta_1 P_t + \beta_2 W_t + \varepsilon_t. \tag{3.2.5}$$

Recall that R_t is the public's relative preference for social policy spending in a given year. P_t is the actual level of policy spending in a year. W_t represents other, exogenous effects on the public's relative preferences.

The validity of any results from estimating this model by OLS is contingent on meeting the Gauss-Markov assumptions and the assumption of stationarity. In thinking about whether the necessary assumptions of OLS regression apply, we would ask ourselves a series of questions. First, is it reasonable to assume that an appropriate model of policy responsiveness is linear in its parameters (A.1)? For example, the public's relative preference for social policy spending may not be a linear function of actual policy spending. However, it may be a linear function of the log of actual policy spending, in which case we would want to use this as our independent variable.

If we have an independent variable in our model with no variation or if we have independent variables that are perfectly collinear (A.2), then any statistical package will pick up the problem and omit one or more variables to resolve the problem. However, we might want to ask ourselves if we have independent variables that are highly collinear. This would not be a concern of bias, but we might want to keep in mind that our standard errors are being increased by it.

We will also want to ask ourselves whether there are any obvious violations of strict exogeneity (A.3). This includes the usual concerns about spurious correlations and omitted-variable bias. It is also a possibility that the data-generating process includes a lag of the dependent variable, which we would then want to include in our data model. We will examine how to address this in Chapter 4. Not controlling for a trend that exists in the data-generating process can also be a violation of exogeneity. It is also a violation of covariance stationarity, which we should remember we are always assuming or must account for in the time series models discussed in this text. We will look at how to test and correct for trending in the next section.

Finally, we will ask ourselves whether or not our errors have constant variance—homoskedasticity—(A.4), no serial correlation (A.5), and are independently and normally distributed (A.6). We will examine how to test and correct for violations of these later in this chapter. We sometimes summarize Assumptions A.4 to A.6 by stating that the ε_t values in an equation such as Equation 3.2.5 are independent and distributed normally with mean 0 and constant variance σ_ε^2, or $\varepsilon_t \sim \text{NID}(0, \sigma_\varepsilon^2)$.

Next, we turn to a scenario in which we have a large number of time points. If we have such data, we can relax the strict-exogeneity assumption and still obtain asymptotic unbiasedness (and consistency). This also has a parallel with cross-sectional data analysis. For cross-sectional data with a large N, it is possible to relax the zero conditional mean assumption to no correlation between the exogenous variables and the errors. With time series data with a large T, it is possible to relax the strict-exogeneity assumption to contemporaneous exogeneity, *if* we can make an additional assumption. As discussed in Chapter 2, that assumption is weak dependence.

Weak dependence (with covariance stationarity) allows us to relax the strict-exogeneity assumption for large-T analysis in demonstrating that OLS is consistent. As T goes to infinity, the sample variances and covariances converge on the population variances and covariances. In this case, β_1 can be estimated as follows:

$$\hat{\beta}_1 = \frac{\widehat{\text{Cov}(Y_t, X_t)}}{\widehat{\text{Var}(X_t)}}. \qquad (3.2.6)$$

We can state the "asymptotic" Gauss-Markov assumptions for time series regression as follows:

AA.1 Linearity and weak dependence

AA.2 Variance in all X and no perfect collinearity

AA.3 Zero conditional mean (contemporaneous exogeneity)

AA.4 Homoskedasticity

AA.5 No serial correlation

Note that Assumptions AA.1 through AA.3 are required for the parameter estimates to be asymptotically unbiased. AA.4 and AA.5 are required to estimate the correct parameter standard errors.

3.3 Testing and Correcting for Trending, Periodicity, and Structural Breaks

The weaker form of stationarity that we assume (covariance stationarity) requires that the mean, variance, and covariances are constant across time. As discussed in Chapter 2, any trending, periodicity, or structural breaks will violate this assumption. It is therefore important to account for any of these processes. We begin with trends. One possibility is a deterministic linear trend, which can be modelled as follows:

$$y_t = \beta_0 + \beta_1 t + \varepsilon_t, \quad t = 1, 2, \ldots \quad (3.3.1)$$

Another possibility is a deterministic quadratic trend, which can be modelled as follows:

$$y_t = \beta_0 + \beta_1 t + \beta_2 t^2 + \varepsilon_t, \quad t = 1, 2, \ldots \quad (3.3.2)$$

We will discuss other types of trends in Chapter 6.

In Chapter 2, we found that we got very different results when we estimated the Canadian public responsiveness model with a trend from what we obtained when we estimated it without a trend. Before beginning that analysis, we might have plotted the dependent variable—the public's relative preference for social policy spending (Figure 3.3).

Figure 3.3 Public's Relative Preference for Social Policy Spending—Canada

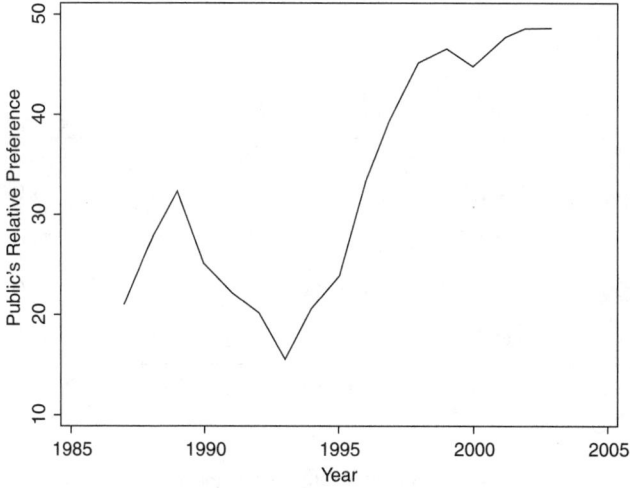

Clearly, the public's relative policy spending preference has been going up over time. However, it has not been going up smoothly. It appears to be exhibiting trending but not linear trending. We may consider the possibility that the public's relative policy spending preference contains a quadratic trend.

$$R_t = \beta_0 + \beta_1 P_t + \beta_2 t + \beta_3 t^2 + \varepsilon_t. \quad (3.3.3)$$

The results from estimating such a model by OLS are given in Table 3.3.

We see that the squared time component in the quadratic trend is significant at the 0.05 significance level. In Chapter 2, we estimated this model with a linear trend. Which should we use? The significance of the squared time variable indicates that the linear trend is insufficient, which suggests that we should use the quadratic trend.[1] Ultimately, we will want to choose the model that ensures the errors are independent and normally distributed with mean 0 and constant variance σ_ε^2: $\varepsilon_t \sim NID(0, \sigma_\varepsilon^2)$. We shall test for this momentarily.

How do we interpret the significance of the quadratic trend terms? In Chapter 2, we noted that the interpretation of a linear trend is that y_t increases or decreases on average by the magnitude of β_2 each time period. A quadratic term means that y_t increases or decreases on average by a decreasing or increasing magnitude, of $\beta_2 + \beta_3(2t)$ each time period. Also noted in Chapter 2, trend terms (linear or quadratic) are controls for violations of covariance stationarity and should only be interpreted as controls. Normally, any further substantive interpretation of these components of the model should be avoided.

Interpreting the coefficient on program spending, for each billion-dollar increase in government spending, the public's relative preference for

Table 3.3 Canadian Public Responsiveness Model

Preference	Coefficient	Standard Error	t Statistic	P Value
Program spending	−0.57	0.10	−5.41	<0.001
Counter	1.47	0.84	1.75	0.106
Counter^2	0.16	0.047	3.29	0.006
Constant	115.55	16.83	6.87	<0.001

NOTE: $R^2 = 0.93$, $T = 16$; T = number of time points.

---

[1] A more advanced method for controlling for complex time trends is the use of regression splines (Durlauf & Blume, 2010; Keele, 2008).

spending decreases by 0.57 of a percentage point. Recall that relative preference is the difference between the percentage of respondents who want more spending and the percentage of respondents who want less.

An alternative to adding a linear or quadratic trend to a regression is to use "detrended" data in the regression. Detrending a series involves regressing each variable in the model on t (or a more complex function of time) and predicting the errors. The predicted errors from each of these regressions form the detrended series for each variable; the trend has been partialled out. Let us consider an example.

In this example, we estimate a public responsiveness model for the United Kingdom. We begin by plotting the dependent variable (Figure 3.4).

As with the Canadian public's relative spending preference, the upward trend in the U.K. public demand for an increase in major social program spending is self-evident. Accordingly, we could include a quadratic time trend in our U.K. model. The results are presented in Table 3.4.

The results indicate that an increase in major social program spending results in a decline in the public's relative preference—this means relatively less demand for spending. For each billion-pound increase in government

Figure 3.4　Relative Preference for Social Program Spending in the United Kingdom

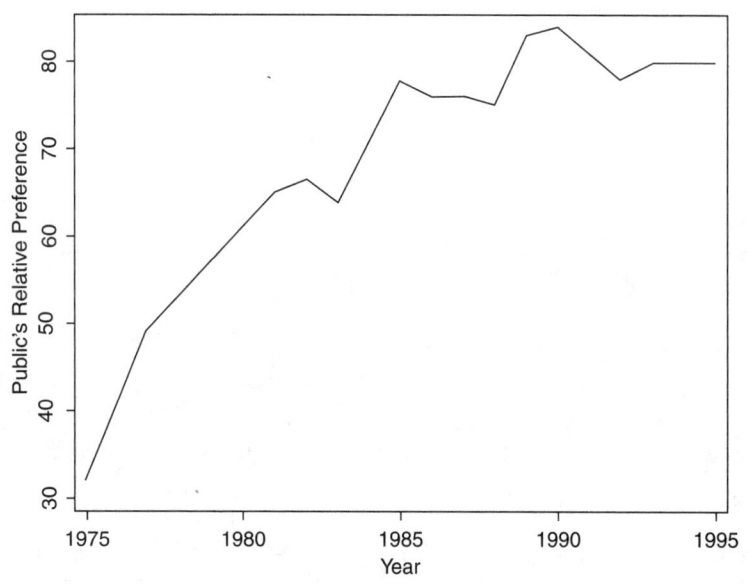

spending, the public's relative preference for spending decreases by 0.98 of a percentage point.

Alternatively, we could detrend the public's relative spending preference and government policy spending variables. We regress the public's relative spending preference on the time trend variables. The results are presented in Table 3.5.

Using these results, we can estimate the errors (the residuals) for this model. This is our detrended public's relative spending preference variable. We then do the same for the government policy spending variable (Table 3.6).

Table 3.4   U.K. Public Responsiveness Model

Preference	Coefficient	Standard Error	t Statistic	P Value
Program spending	−0.98	0.37	−2.62	0.020
Counter	4.09	0.37	10.95	<0.001
Counter^2	−0.05	0.04	−1.17	0.261
Constant	111.52	22.45	4.97	<0.001

NOTE: $R^2 = 0.96$, $T = 18$; T = number of time points.

Table 3.5   U.K. Public's Relative Spending Preference

Preference	Coefficient	Standard Error	t Statistic	P Value
Counter	4.03	0.44	9.14	<0.001
Counter^2	−0.14	0.02	−5.78	<0.001
Constant	52.73	1.61	32.67	<0.001

NOTE: $R^2 = 0.94$, $T = 18$; T = number of time points.

Table 3.6   U.K. Government Policy Spending

Spending	Coefficient	Standard Error	t Statistic	P Value
Counter	0.071	0.26	0.28	0.787
Counter^2	0.097	0.015	6.63	<0.001
Constant	60.12	0.95	63.44	<0.001

NOTE: $R^2 = 0.98$, $T = 22$; T = number of time points.

The residuals from these two estimated models provide us with our two detrended variables. These are plotted in Figure 3.5.

It is clear that the trend is removed from the public's relative spending preference variable, as is the case with the government policy spending variable.

We can now regress the detrended public's relative spending preference variable on the detrended government policy spending variable. We do not include a constant in this model as it has been partialled out along with the trend (Table 3.7).

The results are equivalent to those we get from including the trend variable in the model (c.f. Table 3.4). An advantage of detrending the data (vs. adding

Figure 3.5   Detrended Government Policy Spending and Public Relative Spending Preference

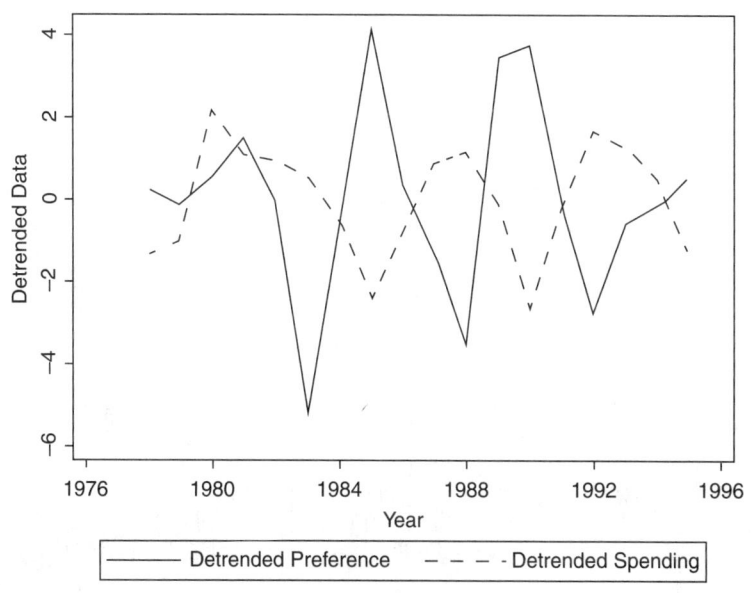

Detrended Preference   – – – Detrended Spending

Table 3.7   U.K. Public Responsiveness Model—Detrended

Preference	Coefficient	Standard Error	t Statistic	P Value
Detrended spending	−0.98	0.34	−2.89	0.010

NOTE: $R^2 = 0.33$, $T = 18$; T = number of time points.

a trend) involves the calculation of goodness of fit. Time series regressions with trend variables tend to have very high R^2 as the trending data are well explained by the trend variable. The R^2 from a regression of the detrended dependent variable on the detrended independent variables better reflects how well the X_ts explain Y_t. Compare the R^2 value in Table 3.7 with that in Table 3.4. The disadvantage of detrending is that the uncertainty in the estimation of the trend is not reflected in subsequent analyses using the detrended data; so the standard errors are a bit too small. This is generally not a big concern.

We turn now to periodicity and structural breaks. As we have seen, even if those factors causing trending are unobserved, we can control for them by directly controlling for the trend. This applies equally to structural breaks and periodicity. We examine this using an example that looks at traffic injuries in the United Kingdom from 1969 to 1984. Consider the monthly data on the number of drivers in the United Kingdom who were killed or seriously injured (KSI) in a traffic accident (Harvey & Durbin, 1986) between January 1969 and December 1984. Figure 3.6 provides a plot of the natural log of KSI.

Figure 3.6   ln(KSI), United Kingdom, From 1969 to 1984

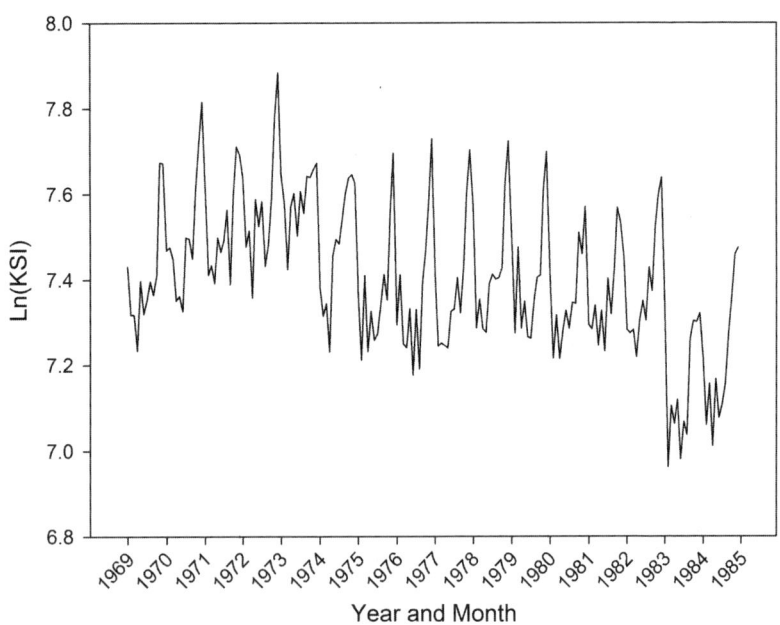

The data appear to exhibit periodicity. Specifically, the data appear to go through a complete cycle each year, with KSIs hitting a peak in winter and a nadir in summer. Periodicity that follows a seasonal pattern is called seasonality. There also appears to have been a sudden downward shift in KSIs in February 1983, when a new seatbelt law was introduced.

We begin by testing for seasonality and a structural break at $t = 170$ (February 1983). We include a time variable to capture any trending and a series of dummy variables for all but one month. We exclude December, so this will be the reference month for the coefficients on the included dummy variables. We also include a variable called "seatbelt law," which is coded "0" before $t = 170$ and "1" on and after $t = 170$. We can use this to test for a structural break in the equilibrium of ln(KSI) at the introduction of the new seatbelt law. Finally, we test for the effect of (the log of) petrol prices (Table 3.8).

Table 3.8   Static Model of ln(KSI), United Kingdom, From 1969 to 1984

ln(KSI)	Coefficient	Standard Error	t Statistic	P Value
Counter	−0.001	0.0001	−5.10	<0.001
January	−0.24	0.029	−8.43	<0.001
February	−0.35	0.029	−12.51	<0.001
March	−0.31	0.029	−11.09	<0.001
April	−0.39	0.029	−13.74	<0.001
May	−0.30	0.028	−10.62	<0.001
June	−0.33	0.028	−11.80	<0.001
July	−0.28	0.028	−10.01	<0.001
August	−0.27	0.028	−9.71	<0.001
September	−0.24	0.028	−8.45	<0.001
October	−0.16	0.028	−5.74	<0.001
November	−0.056	0.028	−1.99	0.049
Seatbelt law	−0.15	0.022	−6.63	<0.001
Petrol—ln(£)	−0.34	0.057	−6.06	<0.001
Constant	6.95	0.14	50.81	<0.001

NOTE: $R^2 = 0.80$, $T = 192$; T = number of time points, KSI = killed or seriously injured.

On examining the results in Table 3.8, there appears to be significant seasonality with fewer KSIs in all months compared with December (especially February through August). This is apparent in the statistical significance (at the 0.05 level) of the monthly dummy variables. Furthermore, an F test of the joint significance of the 11 monthly dummy variables is $F(11, 177) = 34.66$, with a P value <0.001.[2] It also appears that the price of petrol, controlling for seasonality and the new seatbelt law, has a negative effect on the number of KSIs. Finally, the average number of KSIs seems to have dropped considerably following the introduction of the new seatbelt law.

The estimated coefficient for the seatbelt law variable indicates that, on average and holding log of the petrol prices and seasonality constant, there were 15% fewer KSIs per month.[3] The variable captures a level change in the number of KSIs (logged). The seatbelt law is an intervention, and testing for its effect in this manner is a form of intervention analysis. We will return to this more formally in Chapter 5.

3.4 Testing for Serial Correlation, Heteroskedasticity, and Nonnormally Distributed Errors

Before moving on to the topic of testing for serial correlation, let us examine the consequences of violating the assumption of no serial correlation. Consider two separate OLS estimations of a data model

$$y_t = \hat{\beta}_0 + \hat{\beta}_1 x_t + \hat{\mu}_t. \tag{3.4.1}$$

1. On data that have a data-generating process of

$$y_t = \beta_0 + \beta_1 x_t + \varepsilon_t. \tag{3.4.2}$$

2. On data that have a data-generating process of

$$y_t = \beta_0 + \beta_1 x_t + \mu_t, \tag{3.4.3}$$
$$\mu_t = \rho \mu_{t-1} + \varepsilon_t,$$

with ε_t in both cases being independent and normally distributed with mean = 0 and constant variance = σ_ε^2: $\varepsilon_t \sim \text{NID}(0, \sigma_\varepsilon^2)$.

---

[2] For an overview of the F test, see Foster and Christian (2008).

[3] This is based on the following approximation, appropriate when the log of y_t is linearly related to x_t: $\%\Delta y_t \approx (100\beta_1)\Delta x_t$.

Note that μ_t in (Equation 3.4.3) is AR(1), an autoregressive process of order 1, as it contains one lag of itself on the right-hand side, μ_{t-1}. In this equation, we are describing the data-generating process of the error term and so use ρ in place of α_1. This type of process is one possible source of serial correlation in error terms. We will discuss others in this section.

When using the usual estimator of the variance of $\hat{\beta}_1$, we assume that the residuals (errors) are independent—no serial correlation:

$$\mathrm{Var}\left(\hat{\beta}_1\right) = \frac{\sigma^2}{\sum\left(x_i - \overline{x}\right)^2}. \tag{3.4.4}$$

With serial correlation, we cannot make this assumption, and so the calculation must account for this. If serial correlation is the product of first-order autoregression, the correct calculation is

$$\mathrm{Var}\left(\hat{\beta}_1\right) = \left(\frac{\sigma^2}{\sum\left(x_t - \overline{x}\right)^2}\right) + 2\sigma^2\left(\frac{\sum_{t=2}^{T}\left(x_t - \overline{x}\right)\left(x_{t-1} - \overline{x}\right)}{\sum_{t=1}^{T}\left(x_t - \overline{x}\right)^2}\right)\rho. \tag{3.4.5}$$

Note that

$$\left(\frac{\sum_{t=2}^{T}\left(x_t - \overline{x}\right)\left(x_{t-1} - \overline{x}\right)}{\sum_{t=1}^{T}\left(x_t - \overline{x}\right)^2}\right)$$

is the estimated autocorrelation in x_t. This autocorrelation is commonly positive, and the serial correlation (ρ) is usually positive. Therefore, the second term in Equation 3.4.5 will very often be positive. The consequence is that the serial correlation in the errors usually results in larger standard errors, which are not accounted for by the usual estimator. In other words, the usual estimator underestimates the standard errors. Note though, if the autocorrelation in the independent variable is 0, the standard errors will not be misestimated even in the presence of serially correlated errors.

As an example, let us return to the Canadian public responsiveness model. Let us first check whether or not there is autocorrelation in the independent variable P_t (preferences). Specifically, let us check whether or not it is an autoregressive AR(1) process. This is achieved by regressing P_t on a lag of itself, P_{t-1}. As we previously noted that the series might contain a trend, we use detrended data for our test. Accordingly, the regression does not include a constant. This was partialled out in the detrending. The regression results are presented in Table 3.9.

Table 3.9 Canadian Social Program Spending as an AR(1) Process

P_t	Coefficient	Standard Error	t Statistic	P Value
P_{t-1}	0.76	0.17	4.46	0.001

NOTE: $R^2 = 0.59$, $T = 15$; AR = autoregressive, T = number of time points.

Looking at Table 3.9, we can see that P_t does contain autocorrelation. The coefficient on P_{t-1} is statistically significant. This means that if the residuals in our public responsiveness model contained serial correlation, the usual estimate of the standard errors of our model coefficients will be incorrect—quite possibly too small. Given this, we want to be able to test for whether the estimated errors from our data model (Table 3.3 for our relative preference model) are serially correlated or not. If we believe that the potential serial correlation is AR(1), we might proceed by testing the null hypothesis, H_0: $\rho = 0$, in $\mu_t = \rho\mu_{t-1} + \varepsilon_t$, where μ_t is the error from our data model and it is assumed that $\varepsilon_t \sim$ NID$(0, \sigma_\varepsilon^2)$.

If we can assume that we have only strictly exogenous independent variables in the data model, the test is very straightforward. We simply calculate and regress the residuals from the estimated data model on the lagged residuals and use a t test to test the significance of ρ. We can test for AR(q) serial correlation in the same basic manner as for AR(1). AR(1) errors are an example of serial correlation, but not all serial correlation is of this form. For example, μ_t could be a function of μ_{t-2} and μ_{t-1}. This would be an AR(2) process:

$$\mu_t = \rho_1\mu_{t-1} + \rho_2\mu_{t-2} + \varepsilon_t. \qquad (3.4.6)$$

We will discuss higher-order autoregressive processes further in Chapter 5. To test for higher-order serial correlation, we can include q lags of the residuals in the test regression and test for joint significance of the coefficients on the lags of the residuals using an F test (Agresti & Finlay, 2009, chap. 11). However, this approach will only be valid for a relatively large sample size T, because these regressions include one or more lags of the dependent variable (μ_t), which you will remember from Chapter 2 can never be strictly exogenous. We can have contemporaneous exogeneity, $E(\varepsilon_t | \mu_{t-1}) = 0$, but contemporaneous exogeneity is only sufficient with a large sample size, T.

Another test of AR(1) serial correlation is the Durbin-Watson statistic (Bryman, Liao, & Lewis-Beck, 2004; Ostrom, 1990; Durbin & Watson, 1950). The logic of this test is the same as regressing the residuals on a lag of themselves and calculating a t statistic for the resulting parameter. The validity of the Durbin-Watson statistic does not require a large sample size. However, it does still assume strictly exogenous independent variables, and the Durbin-Watson test can produce inconclusive results. Consequently, this test is rarely used anymore.

Alternatively, we can use Durbin's alternative test, also known as Durbin's h (Durbin, 1970). The logic of this test is conceptually equivalent to regressing the residuals on the lagged residuals and all of the independent variables, X_ts. The inclusion of the X_ts controls for the possibility that an included dependent variable is correlated with μ_{t-1}, so we don't need the assumption of strict exogeneity—contemporaneous endogeneity will do as long as we have a large T. Furthermore, the Durbin's h statistic is chi-squared distributed, and so we can calculate a P value to test the null hypothesis of no serial correlation, H_0: $\rho = 0$. Tests for serial correlation assume homoskedasticity of errors; however, Durbin's alternative test can be made robust to heteroskedasticity in most statistical software packages. The downside is that this test is valid asymptotically. Therefore, if we do have a relatively small T, the test is problematic. Increasingly, the next two tests we will discuss are being used in place of the Durbin-Watson or Durbin's alternative test.

The common practice is to use one of two tests. The first is the Portmanteau (Q) test to test the null hypothesis that the residuals form a white noise process (Ljung & Box, 1978). Such a process contains no serial correlation. If $\mu_t = \varepsilon_t$ and $\varepsilon_t \sim \text{NID}(0, \sigma^2)$, μ_t is a white noise process. So we can apply the Q test to the residuals of our estimated data model as a test of serial correlation. This tests that the error autocorrelations are jointly zero, based on the first p autocorrelations, as described in Chapter 2.

$$Q = T(T+2)\sum_{s=1}^{p} \frac{\hat{\rho}_s^2}{T-s},$$
$$H_0: \rho_1, \rho_2, \dots, \rho_p = 0.$$

Recall that we specify p when conducting the test. The Q test tests for more than just autoregressive errors. The white noise test will also test for moving average errors. We have not yet learned about moving average errors, but we will do so in Chapter 4. For now, it suffices to say that it is another source of serial correlation. It should also be noted that we may reject the null hypothesis of a white noise process because of factors other than serial correlation, such as trending errors. We will also address this possibility in Chapter 4.

Importantly, the Q test does not assume strictly exogenous independent variables and does not require a large T.[4]

---

[4] The Q test is valid asymptotically ($T \to \infty$) but it has good small T properties (Harvey, 1993).

The alternative method is to use a test called the Breusch-Godfrey Lagrange Multiplier test (Breusch, 1978; Godfrey, 1978; Wooldridge, 1991, 2006). This tests the null hypothesis of no serial correlation against the alternatives of autoregressive errors up to an order of q and/or moving average errors up to an order of q. We specify q when conducting the test. This test also does not require strictly exogenous independent variables. The Breusch-Godfrey test uses the R^2 value from the regression of the residuals on q lags of themselves and the independent variables. The test statistic is computed as TR^2. It is chi-squared distributed with degrees of freedom equal to q. The Breusch-Godfrey test will have somewhat more power to reject the null hypothesis than the Q test when the null hypothesis is not true (Greene, 2003).

Returning to the Canadian public responsiveness model, we apply the Q test to the residuals. The Q statistic is 15.83 and is chi-squared distributed with degrees of freedom equal to p. In this instance, $p = 6$. We can calculate the corresponding P value for the test: 0.015. The null hypothesis is that the errors are a white noise process. The Q statistic indicates that we can reject the null hypothesis of the errors being a white noise process, suggesting that we may have serial correlation.

We next apply the Breusch-Godfrey test, testing for serial correlation up to an order of 1, then 2, then 3, and all the way up to 6. The results are presented in Table 3.10.

We can reject the null hypothesis of no serial correlation in up to six lags of the errors. Note that this test does not test serial correlation at lag q controlling for serial correlation at lower lags. This means, for example, that if

Table 3.10 Breusch-Godfrey Lagrange Multiplier Test for Autocorrelation—Canada

lags(p)	χ^2	P Value
1	1.76	0.184
2	2.08	0.354
3	5.14	0.162
4	8.98	0.062
5	10.73	0.057
6	13.78	0.032

NOTE: χ^2 =chi-squared statistic.

the residuals are AR(1) serially correlated and we test for second-order serial correlation, we will most likely find that we can reject the null hypothesis of no second-order serial correlation. This is a bit of a problem as it can make it a challenge to figure out at what order the serial correlation exists. In Chapter 5, we will look at how to address this challenge using autocorrelation and partial autocorrelation functions.

As a general rule, it is good practice to test residuals for serial correlation by testing the residuals against the null of a white noise process using the Q test. If higher-order serial correlation of a specific order is suspected, the Breusch-Godfrey test can be used. Higher-order serial correlation of this sort may be present in seasonal data; for example, quarterly data may exhibit 4th-order serial correlation, and monthly data may exhibit 12th-order serial correlation.

In the next section we look at how to correct for serial correlation but first let us look at how to test for heteroskedasticity. As in the cross-sectional case, heteroskedasticity does not cause bias or inconsistency. It may, however, invalidate the standard errors (and therefore the t statistics and F statistics). To test for heteroskedasticity, we can use the Breusch-Pagan test, which involves estimating $\hat{\varepsilon}_t^2 = \delta_0 + \delta_1 x_{1,t} + \cdots + \delta_k x_{k,t} + \delta_k x_{k+1,t} + v_t$ and testing H_0: $\delta_1 = \delta_2 = \cdots = \delta_{k+1} = 0$ with an F statistic (Wooldridge, 2006, pp. 278–281).[5] In other words, we regress the squared residuals on our independent variables and test the null hypothesis that the resulting slope coefficients are jointly equal to 0.[6] The Breusch-Pagan test assumes that the $\hat{\varepsilon}_t$ values are not serially correlated, so serial correlation should be accounted for first. We will see how to do that in the next section of this chapter. The Breusch-Pagan test also assumes that exogeneity is not violated and the functional form of the model is correct (e.g., the relationships that are assumed to be linear really are linear).

We now apply the Breusch-Pagan test to the U.K. public responsiveness model from Table 3.4 to test for heteroskedasticity with respect to the government spending variable. The test statistic is 3.80 and is chi-squared distributed with 3 degrees of freedom (because we are testing for heteroskedasticity with respect to three variables). The P value is 0.28. The null hypothesis is that the errors are homoskedastic. The Breusch-Pagan test indicates that we cannot reject the null hypothesis of constant variance (homoskedasticity).

Finally, we can also test for the skewness and kurtosis of the errors to determine if there is evidence that they are not normally distributed. This is

---

[5] Alternatively, a Lagrange multiplier statistic could be calculated (Koenker, 1981).

[6] The squared errors are proportional to the error variances.

a necessary assumption for the usual methods of inference (again, t and F statistics) when we have a small number of time points on which to estimate our models.

Again, testing the estimated errors from the U.K. public responsiveness model from Table 3.4, the P value for the null hypothesis of no skewness is 0.48, and the P value for the null hypothesis of no kurtosis (relative to the normal) is 0.33. The results indicate that we cannot reject the null of normally distributed errors.

Note that there is no straightforward test for exogeneity. This is an assumption about the data-generating process that we never observe. We can only ever test the exogeneity assumption indirectly by estimating models with different specifications. For example, if we believe that we have an omitted-variable bias problem, we may estimate a model with that variable, or a proxy for it, included and observe the consequences.

3.5 Correcting for Serial Correlation and Heteroskedasticity

We now turn to what we would do if we believed that we have serial correlation. We motivate this by running a U.S. public responsiveness model, including a quadratic trend just as we did in the Canadian and U.K. models.

We next test for serial correlation of the errors by using the Q test to test if the residuals form a white noise process. The Q statistic is 38.36, with a P value of <0.001. We can reject the null hypothesis of a white noise process and suspect that we have serial correlation. We have a number of options available to us for the purpose of correcting the violation of this assumption. As with the tests of serial correlation, we maintain that all

Table 3.11   U.S. Public Responsiveness Model

Preference	Coefficient	Standard Error	t Statistic	P Value
Program spending	−0.23	0.058	−3.93	<0.001
Counter	0.67	0.37	1.79	0.084
Counter^2	0.07	0.022	3.36	0.002
Constant	44.04	7.06	6.24	<0.001

NOTE: $R^2 = 0.67$, $T = 33$; T = number of time points.

Gauss-Markov assumptions hold except no serial correlation. Let us assume that the data-generating process for the errors is an AR(1) process, so it is as follows:

$$y_t = \beta_0 + \beta_1 x_t + \mu_t,$$

$$\mu_t = \rho\mu_{t-1} + \varepsilon_t. \tag{3.5.1}$$

We proceed by estimating Equation 3.5.1 by maximum likelihood. This is unlike the models we have estimated so far, for which we have used OLS. Good introductions to the maximum likelihood approach to model estimation can be found in Fox (2008), Gill (2000), and King (1998).

The results for the maximum likelihood estimation of our U.S. public responsiveness model with AR(1) errors are presented in Table 3.12.

The results include the estimated coefficient of the AR(1) serial correlation. This parameter is statistically significant at the 0.05 significance level. Let us now test the residuals from this model with the Q test. The Q statistic is 11.97, with a P value of 0.61. We cannot reject the null hypothesis of a white noise process.

Another option for accounting for serial correlation is to calculate serial correlation robust standard errors. The idea is to scale the standard errors from an OLS estimation to take serial correlation into account. Newey-West standard errors correct the standard errors estimated by OLS regression (Newey & West, 1987). The error structure is assumed to be

Table 3.12 Maximum Likelihood Estimation of the U.S. Public Responsiveness Model With AR(1) Errors

Preferences	Coefficient	Standard Error	z Statistic	P Value
Program spending	−0.17	0.087	−1.98	0.047
Counter	0.65	0.89	0.73	0.47
Counter^2	0.059	0.43	1.38	0.167
Constant	37.03	10.97	3.37	0.001
ρ	0.66	0.18	5.10	<0.001

NOTE: Log likelihood = −91.352, $T = 33$; T = number of time points, AR = autoregressive.

heteroskedastic and possibly autocorrelated up to some lag.[7] Most statistical software packages have a default procedure to determine how many lags to correct for, accounting for the number of time points in the data. Estimating Newey-West standard errors that correct for first-order serial correlation in the results presented in Table 3.11 produces the results shown in Table 3.13.

Examining Table 3.13, note that the estimated coefficients are just those estimated by OLS (c.f. Table 3.11). It should be noted though that Newey-West standard errors can be poorly behaved with a low T (Wooldridge, 2006). Also, note that Newey-West standard errors correct for heteroskedasticity. If we are not using Newey-West standard errors and we believe we have homoskedasticity, we must calculate heteroskedasticity robust standard errors. Heteroskedastic robust errors are easily computed by most statistical packages.

Let us finish this chapter with an example that puts together everything we have learned in this chapter. In this example, we examine the relative legislative success of minority and majority parliamentary governments. We will examine the Canadian federal parliament. Legislative success is the ability of a government to pass the legislation that it deems important. The expectations are that the legislative success of a minority government will generally be lower than that of a majority government and the legislative success of a minority government will increase as its popularity in the polls published in the media increases.

The measure of legislative success used is the proportion of bills the government moved past second reading in the House of Commons that received Royal Assent in each session. Bills that do not pass second reading

Table 3.13 Newey-West Standard Errors for the U.S. Public Responsiveness Model

Preference	Coefficient	Standard Error	t Statistic	P Value
Spending	−0.23	0.062	−3.67	<0.001
Counter	0.67	0.45	1.50	0.144
Counter^2	0.074	0.021	3.61	<0.001
Constant	44.04	7.06	6.24	<0.001

NOTE: $R^2 = 0.36$, $T = 33$; T = number of time points.

---

[7] See Wooldridge (2012) for a straightforward exposition.

are not considered.[8] The measure of the popularity of the government is based on all published results from polls asking respondents their vote intention. A typical vote intention question is "If a federal election were held tomorrow, which party would you vote for?" The measure of popularity is the average share of vote intention the governing party received in all polls published over each session.

As for whether a government is a minority or a majority government, a minority government is simply defined as any government in which the governing party has less than 50% + 1 of the seats in the House of Commons. The minority variable is coded "1" for minority governments and "0" for majority governments. We also control for whether the governing party was 1 = *the Liberals* or 0 = *the (Progressive) Conservatives*.

Our data include these measures for the period extending from the beginning of the 24th to the end of the 40th Parliament. This includes 42 sessions of Parliament and spans the temporal period 1958 to 2008. The temporal unit of analysis is the parliamentary session. Using these measures and a "government popularity × minority" interaction variable, we use OLS regression to estimate a model of legislative success testing the effects of minority versus majority government, government popularity, and their interaction and a control for the party in government. Before doing so, we plot the dependent variable, "legislative success" (Figure 3.7).

The plot of legislative success does not reveal any periodicity or structural breaks. If it did contain either of these processes, they would be a violation of stationarity, and we would have to control for them. There are possibly two outliers. One of which (the 31st Parliament) was a result of a quickly defeated minority government. The inclusion of a minority variable in the model may account for this outlier. The other outlier (the third session of the 37th Parliament) reflects the last session of a long-standing prime minister (Jean Chrétien) and a high degree of infighting within the governing Liberal party over whether or not Chrétien should resign. These outliers might cause issues with homoskedasticity and with the distribution of the errors. It is also possible that there is a systematic decline in legislative success in the later parliamentary sessions. This would be a nonlinear trend that could be modelled with a quadratic function of time. It is possible that this is just an end effect; that is, there appears to be a trend at the end, but if we were able to continue observing this series further in time, we would see that it is just a temporary downward shift (Tufte, 2001). To test the possibility that the series exhibits a nonlinear trend, we regress legislative success

---

[8] Bills were selected in this way because governing parties will sometimes introduce bills that they do not intend to pursue. There are a number of reasons why a governing party may do this. However, if the government pursues a bill past second reading in the House of Commons, one can be relatively certain that it is serious about passing it into law.

on time and time-squared variables. The results (Table 3.14) suggest that no such trending exists as the coefficients on both variables are not statistically significant at the 0.05 significance level.

Satisfied that our results are not going to be biased by trending, periodicity, or structural breaks, we estimate our legislative success model. The results are presented in Table 3.15.

Figure 3.7 Legislative Success of the Canadian Federal Government

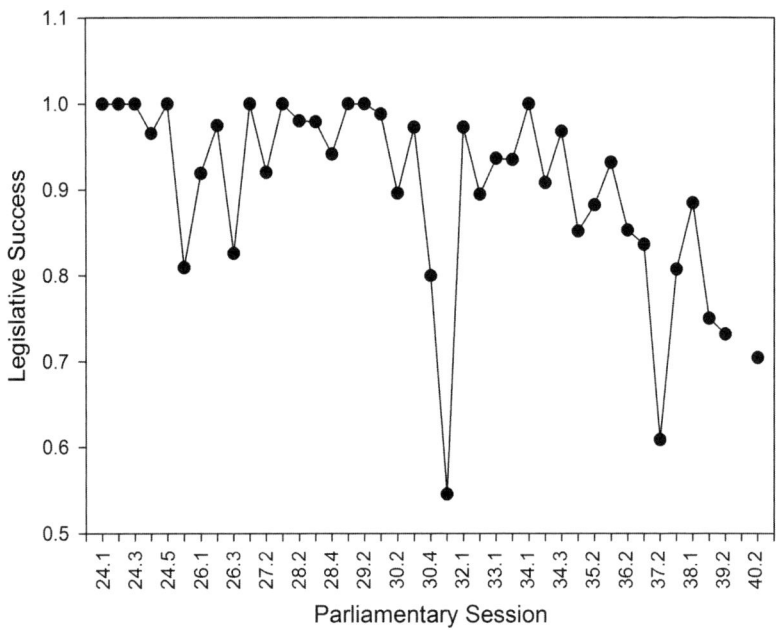

Table 3.14 Test for Trending in Legislative Success

Legislative Success	Coefficient	Standard Error	t Statistic	P Value
Trend	0.0026	0.0051	0.50	0.622
t^2	−0.00019	0.00012	−1.56	0.128
Constant	0.95	0.046	20.69	<0.001

NOTE: $R^2 = 0.34$, $T = 40$; T = number of time points.

Table 3.15 Canadian Legislative Success Model

Legislative Success	Coefficient	Standard Error	t Statistic	P Value
Governing party	−0.02	0.0384	−0.52	0.603
Government popularity	−0.00016	0.0021	−0.08	0.939
Minority government	−0.74	0.21	−3.46	0.001
Govpop × Min	0.017	0.0052	3.25	0.003
Constant	0.94	0.089	10.56	<0.001

NOTE: $R^2 = 0.30$, $T = 40$; T = number of time points.

The estimated coefficient for government popularity is the effect of popularity on the legislative success of majority governments. It is not statistically significant at a 0.05 significance level. This suggests that the legislative success of majority governments is not affected by their standing in the polls. The effect of government popularity on the success of minority governments is the sum of the coefficients on government popularity and the government popularity × minority interaction. The sum of these coefficients is 0.017, and an F test of the hypothesis that the sum of these coefficients is 0 gives us a statistic of $F(1, 35) = 10.57$. The corresponding P value is 0.0025. We can reject the null hypothesis that the sum of the coefficients is 0. It would appear that the legislative success of a minority government is improved as its popularity in the polls increases. However, as this chapter has discussed, these results are contingent on certain assumptions.

At this point, we might want to test the residuals for serial correlation. We do this here by conducting an overall test of white noise on the residuals. We use the Q test for white noise. It gives us a Q statistic of 14.63. This is chi-squared distributed with 18 degrees of freedom and a P value of 0.69. We cannot reject the null hypothesis that the residuals are a white noise process.

Next, we might want to test for heteroskedasticity. The presence of heteroskedasticity would not bias the coefficients but would result in the estimation of incorrect standard errors. We test for heteroskedasticity with respect to the minority, government popularity, and governing party variables. The Breusch-Pagan test for heteroskedasticity gives us a statistic of 3.40, which is chi-squared distributed with 4 degrees of freedom. The corresponding P value is 0.49, which means that we cannot reject the null hypothesis of homoskedasticity.

Finally, we can test whether the residuals deviate from a normal distribution. Again, this is not a necessary condition for unbiased coefficients, but it is necessary for the use of t and F statistics as tests of inference when we have a small number of time points. The tests of the null hypothesis of no skewness and of the null hypothesis of no kurtosis (relative to the normal) have P values of 0.004 and 0.032, respectively. This suggests that we can reject the null hypothesis of normally distributed residuals. With 40 time points, we may not be concerned about the deviations from normality. If we were, we would want to investigate the source of the violation.

Regardless of the outcomes of these tests, it is always a good idea to plot the residuals (Figure 3.8). If we observe any systematic pattern in the plot of the residuals, it could be an indication of a violation of an important assumption, such as no trending, no serial correlation, or normally distributed errors. In this way, the residuals plot can be a good visual confirmation of the statistical tests.

The plot reveals that the primary reason why the residuals are not normally distributed is the large negative residual for the previously identified third session of the 37th Parliament. At this point, we might want to think

Figure 3.8 Residuals From the Canadian Legislative Success Model

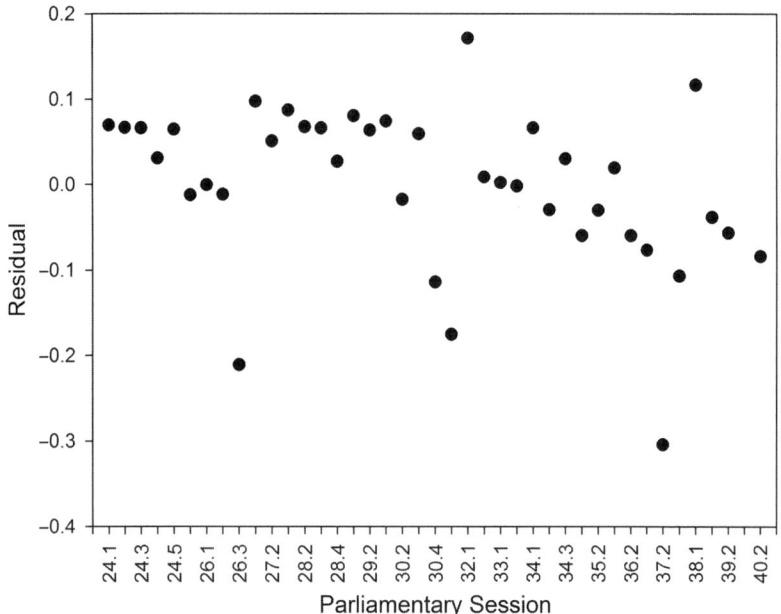

about whether we have excluded an important variable that might explain outliers (e.g., the level of infighting within the governing party) and/or test whether or not these outliers have much leverage on the results. Procedures to do this are exactly as they are in a cross-sectional analysis (e.g., see Andersen, 2008).

Summary

In this chapter, you have been introduced to static and FDL models, and the assumptions necessary to estimate them. In the next chapter, we move on to dynamic time series models and discuss the complications these introduce for unbiased estimation.

CHAPTER 4: INTRODUCING DYNAMIC TIME SERIES MODELS

In this chapter, we continue the discussion of modelling time series data with the introduction of dynamic models. These models include the autoregressive distributed lag (ADL), lagged dependent variable (LDV), autoregressive conditional heteroskedasticity (ARCH), and moving average (MA) models. As the conditions necessary for these dynamic models to meet the assumption of covariance stationarity differ from those of static models, these conditions are discussed.

4.1 Autoregressive Distributed Lag Models

In Chapter 2, we discussed the following process:

$$y_t = \alpha_1 y_{t-1} + \varepsilon_t, \tag{4.1.1}$$

with $\varepsilon_t \sim \text{NID}(0, \sigma_\varepsilon^2)$. In other words, the ε_t values are independent and distributed normally with mean 0 and variance σ_ε^2. If $|\alpha_1| < 1$, this is an autoregressive process of order 1, AR(1). It contains one lag of itself on the right-hand side, y_{t-1}.

An autoregressive process is described by including lags (or just a single lag) of the dependent variable on the right-hand side as explanatory/independent variables. We may also include a constant, α_0, in Equation 4.1.1:

$$y_t = \alpha_0 + \alpha_1 y_{t-1} + \varepsilon_t. \tag{4.1.2}$$

Note that we are using α_0 as the constant, rather than β_0, in this model. This constant implies that the process y_t has a nonzero mean. In the static process, $y_t = \beta_0 + \varepsilon_t$, the constant is the unconditional mean or expected value of y_t. This can also be interpreted as the equilibrium of the process—the value to which the process reverts if the stochastic component, ε_t, and the deterministic exogenous variables x_t (if there are any) are set to 0. It is the long-run mean (expected value) of the process. In the AR(1) process, the equilibrium of the process is:

$$\beta_0 = \frac{\alpha_0}{1 - \alpha_1}. \tag{4.1.3}$$

This again is the value to which the process eventually returns if the stochastic and exogenous variables are set to 0. This may not have a practical interpretation. It is the long-run expected value of y_t. More precisely, it is the unconditional expected value of y_t: $E(y_t)$. In a dynamic process, the *conditional* expected value, the mean of y_t conditioning on previous values of y_{t-1}, is

$$E(y_t| y_{t-1}) = \alpha_0 + \alpha_1 y_{t-1}. \tag{4.1.4}$$

Note that it is the unconditional expected value that we assume to be constant in order to meet the conditions of covariance stationarity. Unless the dynamic process begins exactly in equilibrium and remains there, the conditional expected value will change over time.

With this in mind, we can define a dynamic time series model called an autoregressive distributed lag model (ADL). Such a model contains lags of the dependent and independent variables on the right-hand side. In the ADL model, it is the inclusion of a lag or lags of the dependent variable on the right-hand side that makes the model dynamic. The ADL model is a combination of an autoregressive process and a finite distributed lag (FDL) model. An ADL(1,1) model contains one lag of the dependent variable and one lag of the independent variable (it also includes a constant and the nonlagged independent variable):

$$y_t = \alpha_0 + \alpha_1 y_{t-1} + \beta_1 x_t + \beta_2 x_{t-1} + \varepsilon_t,$$

$$\varepsilon_t \sim \text{NID}\left(0, \sigma_\varepsilon^2\right). \tag{4.1.5}$$

More generally, an ADL(p,m) model is an autoregressive process of order p (with p lags of the dependent variable included as independent variables), and it includes m lags of the other independent variable(s), in addition to the nonlagged independent variable(s):

$$y_t = \alpha_0 + \sum_{j=1}^{p} \alpha_j y_{t-j} + \beta_1 x_t + \sum_{i=1}^{m} \beta_{i+1} x_{t-i} + \varepsilon_t. \tag{4.1.6}$$

Interpreting the coefficients on independent variables in dynamic models is somewhat complicated by the inclusion of the lagged dependent variable. From an ADL(p,m) model, the short-run effect and the long-run effect of the independent variables, and the process equilibrium can be calculated. For the ADL(1,m) model, β_1 is the immediate (short-run) effect of a one-unit change in x_t. The long-run effect of a permanent one-unit change is

$$\frac{\beta_1 + \sum_{i=1}^{m} \beta_{i+1}}{1 - \alpha_1}. \tag{4.1.7}$$

The reason why the short-run and long-run effects of an exogenous variable are different is not immediately obvious. This will be explored further in Section 4.6, but in the meantime, suppose we have an ADL(1,1) model, as written in Equation 4.1.5. Also, suppose we observe y_t up to time $t = \tau$. The conditional expected value of y_t is

$$E(y_t) = \alpha_0 + \alpha_1 y_{t-1} + \beta_1 x_t + \beta_2 x_{t-1}. \tag{4.1.8}$$

We can predict the expected value of y_t at time $t = \tau + 1$ using our observations of y_t up to time $t = \tau$ and our observation of x_t at $t = \tau + 1$. Such a forward prediction is a forecast of $y_{t = \tau + 1}$. We denote such a forecast as

$$\hat{E}\left(y_{t=\tau+1|\tau}\right) = \alpha_0 + \alpha_1 y_\tau + \beta_1 x_{\tau+1} + \beta_2 x_\tau. \tag{4.1.9}$$

The term on the left-hand side reads as the estimated (conditional) expected value of y_t at time $y_{t = \tau+1}$ using observations of y_t up to and including $t = \tau$. Suppose x_t increases by one unit at $\tau + 1$. Since the difference between x_τ and $x_{\tau+1}$ is 1, the effect of the one-unit increase in x_t at $\tau + 1$ on y_t at $\tau + 1$ is a β_1 increase. We can next forecast the value of y_t at time $t = \tau + 2$.

$$\hat{E}\left(y_{t=\tau+2|\tau}\right) = \alpha_0 + \alpha_1 y_{\tau+1} + \beta_1 x_{\tau+2} + \beta_2 x_{\tau+1}.$$

We do not know $y_{\tau + 1}$, but we can plug our forecast of $y_{\tau+1}$ into the above:

$$\hat{E}\left(y_{t=\tau+2|\tau}\right) = \alpha_0 + \alpha_1 \left(\alpha_0 + \alpha_1 y_\tau + \beta_1 x_{\tau+1} + \beta_2 x_\tau\right) + \beta_1 x_{\tau+2} + \beta_2 x_{\tau+1}$$

$$= \alpha_0 + \alpha_1 \alpha_0 + \alpha_1^2 y_\tau + \beta_1 x_{\tau+2} + \left(\alpha_1 \beta_1 + \beta_2\right) x_{\tau+1} + \alpha_1 \beta_2 x_\tau. \tag{4.1.10}$$

The effect of the one-unit increase in x_t at $\tau + 1$ (which remains at $\tau + 2$) on y_t at $\tau + 2$ is the sum of the coefficients on $x_{\tau+1}$ and $x_{\tau+2}$:

$$\beta_1 + \alpha_1 \beta_1 + \beta_2 = \beta_1\left(1 + \alpha_1\right) + \beta_2. \tag{4.1.11}$$

Similarly, forecasting the value of y_t at time $t = \tau + 3$,

$$\hat{E}\left(y_{t=\tau+3|\tau}\right) = \alpha_0 + \alpha_1 \alpha_0 + \alpha_1^2 y_{\tau+1} + \beta_1 x_{\tau+3} + \left(\alpha_1 \beta_1 + \beta_2\right) x_{\tau+2} + \alpha_1 \beta_2 x_{\tau+1}$$

$$= \alpha_0 + \alpha_1 \alpha_0 + \alpha_1^2 \alpha_0 + \alpha_1^3 y_\tau + \beta_1 x_{\tau+3} + \left(\alpha_1 \beta_1 + \beta_2\right) x_{\tau+2}$$

$$+ \left(\alpha_1^2 \beta_1 + \alpha_1 \beta_2\right) x_{\tau+1} + \alpha_1^2 \beta_2 x_\tau. \tag{4.1.12}$$

The effect of the one-unit increase in x_t at $\tau + 1$ on y_t at $\tau + 3$ is the sum of the coefficients on $x_{\tau+1}$, $x_{\tau+2}$, and $x_{\tau+3}$:

$$\beta_1\left(1 + \alpha_1 + \alpha_1^2\right) + \beta_2\left(1 + \alpha_1\right). \tag{4.1.13}$$

We can continue forecasting as long as we like, and we will find that the effect of the one-unit increase in x_t at $\tau + 1$ on y_t at $\tau + m$ is

$$\beta_1\left(1 + \alpha_1 + \alpha_1^2 + \cdots + \alpha_1^{t+m}\right) + \beta_2\left(1 + \alpha_1 + \alpha_1^2 + \cdots + \alpha_1^{t+m-1}\right). \tag{4.1.14}$$

If we assume that $|\alpha_1| < 1$, an assumption we will discuss in Section 4.6, then the effect of the one-unit increase in x_t at $\tau + 1$ on y_t increases at a diminishing rate as t increases. If m is a very large number of time points, Equation 4.1.14 converges to

$$\frac{\beta_1}{1 - \alpha_1} + \frac{\beta_2}{1 - \alpha_1} = \frac{\beta_1 + \beta_2}{1 - \alpha_1}.$$

This is the long-run effect of a one-unit increase in x_t, as expressed in Equation 4.1.7.

Let us now return to the example of the U.S. public responsiveness model from Chapter 3. We had found that we had a problem with autoregressive errors, which we addressed by estimating a static model with AR(1) errors. As an alternative, we can run an ADL model. For this model, we will not be able to make the assumption of strict exogeneity. A lagged dependent variable is by definition not strictly exogenous. Strict exogeneity here requires that

$$E\left(\varepsilon_t \mid y_{t-1+h}\right) = 0, \forall h. \tag{4.1.15}$$

But clearly, ε_t is not independent from y_t: $E\left(\varepsilon_t \mid y_{t-1+h}\right) \neq 0, h = 1$. The best we can do with a lagged dependent variable is the following:

$$E\left(\varepsilon_t \mid y_{t-1+h}\right) = 0, \ h \leq 0. \tag{4.1.16}$$

This is called sequential exogeneity. This condition will be met if the values of ε_t are independent from current and past values of the exogenous variables and past values of the lagged dependent variable. Sometimes this condition will be expressed by saying that the variables in the model are *predetermined*. Sequential exogeneity meets the criteria for contemporaneous exogeneity but not strict exogeneity, and so an ADL model requires a sufficient number of time points to allow us to rely on the assumptions of ordinary least squares (OLS) estimator consistency (see Section 3.2 in Chapter 3).

Let us estimate an ADL(1,1) data model, as described in Equation 4.1.5 (Table 4.1). We include the trend we previously determined we needed in Chapter 3.[1]

Let us now test the residuals from this model with the Q test. The Q statistic is 11.97, with a P value of 0.61. We cannot reject the null hypothesis of a white noise process. It appears as though we have successfully taken care of the serial correlation found in the residuals of the static model using the ADL(1,1) model.

Based on these results, we can calculate the estimated short-run and long-run effects of social program spending and the equilibrium of relative public preference. The estimated immediate short-run effect of a one-unit change in social program spending is statistically significant (at the 0.05 significance level): −0.151. A billion-dollar increase in spending immediately (within the year) reduces the relative preference for a spending increase by 0.15 percentage points. The estimated long-run effect of a permanent one-unit change is

$$\frac{\hat{\beta}_1 + \hat{\beta}_2}{1 - \hat{\alpha}_1} = \frac{-0.151 - 0.035}{1 - 0.57} = -0.43.$$

Table 4.1   U.S. Public Responsiveness Model—ADL(1,1) Model

Preference	Coefficient	Standard Error	t Statistic	P Value
L1. Preference	0.57	0.14	4.15	<0.001
Spending	−0.151	0.067	−2.27	0.032
L1. Spending	−0.035	0.082	−0.42	0.676
Trend^2	0.068	0.018	3.76	0.001
Constant	31.05	8.047	3.86	0.001

NOTE: $R^2 = 0.84$, $T = 32$; T = number of time points, ADL = autoregressive distributed lag, L1 = first lag.

---

[1] Note that we include the squared trend term without the linear component. This follows the modelling choice made by Soroka and Wlezien (2010).

A billion-dollar increase in spending has a long-run effect of decreasing the relative preference for a spending increase by 0.43 percentage points.

We can test the statistical significance of the estimated long-run effect using a Wald test (Wooldridge, 2006). We use a chi-squared statistic with 1 degree of freedom. It is 11.88, and the corresponding P value is 0.001. It appears that the long-run effect is statistically significant at the usual 0.05 significance level. The estimated equilibrium of the relative public spending preference is

$$\frac{\hat{\alpha}_0}{1 - \hat{\alpha}_1} = \frac{31.05}{1 - 0.57} = 32.93.$$

In the static model with AR(1) errors from Chapter 3, the short-run effect was the long-run effect. A billion-dollar increase in spending reduced the relative preference for spending by 0.17 percentage points (Table 3.12). Substantively, this tells a different story from that of the ADL(1,1) model. In Section 4.3, we shall discuss in detail the criteria we should use in determining whether to use a dynamic model with a lagged dependent variable or a static model with autoregressive errors.

Next, let us also look at the example of economic popularity in West Germany, using the vote intention data introduced in Chapter 1. Consider the monthly vote intention and economic data for the Kohl government from October 1982 till September 1998; the economic variables are gross domestic product (GDP), unemployment, and inflation.

We estimate the ADL(1,1) model, regressing vote intention (percentage of respondents from a monthly poll indicating that they would vote for the Kohl government in an election) on a lag of vote intention, the three economic variables, lags of the three economic variables, and the trend variable (Table 4.2). It is not necessary to add a lag of t as it would just be a trend variable itself.

$$\text{vote}_t = \alpha_0 + \alpha_1 \text{vote}_{t-1} + \beta_1 \text{GDP}_t + \beta_2 \text{inf}_t + \beta_3 \text{unemp}_t$$

$$+ \beta_4 \text{GDP}_{t-1} + \beta_5 \text{inf}_{t-1} + \beta_6 \text{unemp}_{t-1} + \beta_7 t + \varepsilon_t. \quad (4.1.17)$$

We now test the residuals for serial correlation using the Q test for white noise. The Q statistic is 14.52. The corresponding P value is 0.91. We fail to reject the null hypothesis of white noise.

None of the coefficients on the economic variables on their own appear to be significant, but the long-run effect of the one-unit increase in inflation is

$$\frac{\beta_1 + \beta_2}{1 - \alpha_1} = \frac{0.320 - 2.30}{1 - 0.38} = -3.19.$$

Table 4.2 Economic Popularity in West Germany—ADL(1,1) Model

	Coefficient	Standard Error	t Statistic	P Value
L1. Vote	0.38	0.14	2.75	0.009
GDP	−0.53	0.44	−1.2	0.236
Unemployment	−5.08	5.56	−0.91	0.366
Inflation	0.32	1.27	0.25	0.802
L1. GDP	−0.010	0.43	−0.02	0.981
L1. Unemployment	5.92	5.08	1.17	0.25
L1. Inflation	−2.30	1.31	−1.75	0.087
Trend	−0.23	0.083	−2.77	0.008
Constant	35.22	18.38	1.92	0.062

NOTE: $R^2 = 0.54$, $T = 51$; T = number of time points, ADL = autoregressive distributed lag, GDP = gross domestic product, L1 = first lag.

The chi-squared statistic with 1 degree of freedom is 7.41, and the corresponding P value is 0.007. It appears as though inflation had a significant and negative effect on vote intention for the Kohl government.

To get an understanding of the relationship between dynamic and static models, let us examine what would happen if we have a time series with an ADL(1,1) data-generating process,

$$y_t = \alpha_0 + \alpha_1 y_{t-1} + \beta_1 x_t + \beta_2 x_{t-1} + \varepsilon_t \qquad (4.1.18)$$

with $\varepsilon_t \sim \text{NID}\left(0, \sigma_\varepsilon^2\right)$, but we estimated it with a static model:

$$y_t = \tilde{\alpha}_0 + \tilde{\beta}_1 x_t + \tilde{\mu}_t \qquad (4.1.19)$$

The data-generating process for y_t (Equation 4.1.18) could be reformulated as

$$y_t = \alpha_0 + \beta_1 x_t + \mu_t,$$

$$\mu_t = \alpha_1 y_{t-1} + \beta_2 x_{t-1} + \varepsilon_t \qquad (4.1.20)$$

with $\varepsilon_t \sim \text{NID}\left(0, \sigma_\varepsilon^2\right)$. Therefore, our data model is

$$y_t = \tilde{\alpha}_0 + \tilde{\beta}_1 x_t + \tilde{\mu}_t,$$

$$\mu_t = \alpha_1 y_{t-1} + \beta_2 x_{t-1} + \varepsilon_t. \qquad (4.1.21)$$

Note that we are suggesting that a misspecified model has resulted in a particular data-generating process for the residuals. From Equation 4.1.20,

$$y_{t-1} = \alpha_0 + \beta_1 x_{t-1} + \mu_{t-1}. \qquad (4.1.22)$$

Substituting Equation 4.1.22 for y_{t-1} in the expression for the residuals of Equation 4.1.21,

$$\mu_t = \alpha_1 \left(\alpha_0 + \beta_1 x_{t-1} + \mu_{t-1} \right) + \beta_2 x_{t-1} + \varepsilon_t,$$

$$\mu_t = \alpha_1 \alpha_0 + \left(\alpha_1 \beta_1 + \beta_2 \right) x_{t-1} + \alpha_1 \mu_{t-1} + \varepsilon_t. \qquad (4.1.23)$$

This data-generating process for the residuals is first-order autoregressive with autoregressive parameter α_1. What this tells us is that if we estimate a static model and find that the residuals are autocorrelated, one potential reason for this is that we should have included a lag of the dependent variable in the model.

Also, note the consequence of excluding the lag of the independent variable for the assumption of exogeneity. We do not meet the conditions of sequential exogeneity. The first lag of the independent variable included in Equation 4.1.21, $x_t = x_{t-1}$, is clearly correlated with the error term, which is itself a function of x_{t-1} by Equation 4.1.23.

To further understand the relationship between the static model with AR(1) serially correlated errors from Chapter 3 and the ADL(1,1) model, consider the following. The static model with AR(1) errors is

$$y_t = \beta_0 + \beta_1 x_t + \mu_t. \qquad (4.1.24)$$

$$\mu_t = \rho \mu_{t-1} + \varepsilon_t. \qquad (4.1.25)$$

From Equation 4.1.24,

$$\mu_t = y_t - \beta_0 - \beta_1 x_t.$$

And therefore,

$$\mu_{t-1} = y_{t-1} - \beta_0 - \beta_1 x_{t-1}. \tag{4.1.26}$$

Substituting Equation 4.1.25 into Equation 4.1.24,

$$y_t = \beta_0 + \beta_1 x_t + \rho \mu_{t-1} + \varepsilon_t. \tag{4.1.27}$$

Substituting Equation 4.1.26 into Equation 4.1.27,

$$y_t = \beta_0 + \beta_1 x_t + \rho \left(y_{t-1} - \beta_0 - \beta_1 x_{t-1} \right) + \varepsilon_t.$$

And rearranging terms,

$$y_t = \left(1 - \rho\right)\beta_0 + \rho y_{t-1} + \beta_1 x_t - \rho \beta_1 x_{t-1} + \varepsilon_t. \tag{4.1.28}$$

Define[2] $\left(1 - \rho\right)\beta_0 \equiv \alpha_0$; $\rho \equiv \alpha_1$, and $-\rho\beta_1 \equiv \beta_2$, and our static model with AR(1) errors has now become an ADL(1,1) model:

$$y_t = \alpha_0 + \alpha_1 y_{t-1} + \beta_1 x_t + \beta_2 x_{t-1} + \varepsilon_t,$$

with the restriction that $\beta_2 = -\alpha_1 \beta_1$. We might ask, "What does this restriction mean?" Recall that the long-run effect for an ADL(1,1) model is

$$\frac{\beta_1 + \beta_2}{1 - \alpha_1}.$$

Substituting our restriction $\beta_2 = -\alpha_1 \beta_1$ into the calculation for the long-run effect,

$$\frac{\beta_1 + \beta_2}{1 - \alpha_1} = \frac{\beta_1 - \alpha_1 \beta_1}{1 - \alpha_1} = \frac{\beta_1 \left(1 - \alpha_1\right)}{1 - \alpha_1} = \beta_1.$$

---

[2] Note that we are are assuming that $E(\varepsilon_t \mid x_t) = 0$.

The long-run effect is exactly the same as the short-run effect. This restricted ADL is equivalent to the static model with autoregressive errors, so it is not surprising that the short-run effect is the long-run effect.

We will now consider an alternative restriction to the ADL and in doing so introduce the LDV model.

4.2 Lagged Dependent Variable Models (or Partial Adjustment Models)

If we are willing to assume that $\beta_2 = 0$ in Equation 4.1.18, we have what is called an LDV model, also known as a *partial adjustment* model (Hendry, 2003):

$$y_t = \alpha_0 + \alpha_1 y_{t-1} + \beta_1 x_t + \varepsilon_t. \qquad (4.2.1)$$

An LDV model has no lagged independent variables. Like the ADL model, the LDV model is a useful approach for dealing with serially correlated errors, but they both assume fundamentally different time series processes from the static model with autoregressive errors (unless the ADL is appropriately restricted). We could instead assume that $\beta_1 = 0$ in Equation 4.1.18. This model assumes that the independent variable takes one time period to have an effect. This is known as a *dead start* model (Hendry, 2003).

Returning to our U.S. public responsiveness model, we might note that the lag of the independent variable "social policy spending" is not statistically significant. Based on this information, we might try an LDV model—the difference is the exclusion of lags of the independent variables. Alternatively, we may have hypothesized from the beginning, based on theoretical expectations of the data-generating process, that the LDV model was the most appropriate. In either case, we would estimate Equation 4.2.1 (Table 4.3).

Table 4.3 U.S. Public Responsiveness Model, Social Programs—LDV Model

Preference	Coefficient	Standard Error	t Statistic	P Value
L1.Preference	0.61	0.11	5.77	<0.001
Spending	−0.17	0.043	−4.04	<0.001
Counter^2	0.064	0.015	4.32	<0.001
Constant	28.85	6.05	4.77	<0.001

NOTE: $R^2 = 0.84$, $T = 32$; T = number of time points, LDV = lagged dependent variable, L1 = first lag.

We can test whether or not the estimated errors (the residuals) from our model follow a white noise process. The Q statistic is 12.55 and is chi-squared distributed with 14 degrees of freedom. The corresponding P value is 0.562. We cannot reject the null hypothesis of white noise for the errors. It appears that we do not have serially correlated errors. Recall that we did reject the null hypothesis of white noise for the errors from the estimated static model. We can also test the errors for normality. The P value for the null hypothesis of no skewness is .11, and the P value for the null hypothesis of no kurtosis (relative to the normal) is 0.67. There is no evidence to suggest that we can reject the null hypothesis of skewness or kurtosis different from a normal distribution. Therefore, we cannot reject the null hypothesis of normally distributed errors.

We can estimate the short-run and long-run effects for the LDV model just as we did with the ADL(1,1) model, except that β_2 is now equal to 0. The short-run effect of increasing social policy spending by \$1 billion is to reduce the relative preference for a spending increase by 0.17 of a percentage point. This is statistically significant at the 0.05 significance level. The long-run effect of increasing social policy spending is

$$\frac{\beta_1}{1-\alpha_1} = \frac{-0.17}{1-0.61} = -0.44.$$

The long-run effect of increasing social policy spending by \$1 billion is to reduce the relative preference for a spending increase by 0.44 of a percentage point. The chi-squared test statistic is 10.50, and the corresponding P value is 0.001. This effect is significant at the 0.05 significance level.

4.3 Advice on Model Selection

We have now discussed enough modeling options to raise the issue of how to choose one model over another. We have already discussed the importance of including nonstationary elements, such as trends, periodicity, and structural breaks, in our models. But other issues will have come to your mind as you were introduced to the models of this and the previous chapter.

When we were discussing the ADL(1,1) and LDV models, you may have asked yourself, "Why did we settle on a model with one lag of the dependent variable instead of two or three?" The answer to this question can be found in the structure of the autocorrelations of the errors. We will talk about this further in Chapter 5 when we discuss model selection and testing. For the time being, it is worth noting that a time series model is considered a good fit to the data if the resulting residuals are a white noise process (Li, 2004). Therefore, the lag structure of a model is selected to produce such residuals.

As we will see in Chapter 5, there are often multiple possible lag structures that produce this result, and so additional requirements are applied in model selection. In Chapter 5, we will discuss the requirements used by the Box-Jenkins approach to model selection. However, another approach has emerged more recently within the field of time series analysis. This is the general-to-specific approach (Hendry, 2003).

We have seen that the LDV and the static (and FDL) models are restricted forms of the ADL model. In Chapter 6, we will see that the standard form of a model known as the error correction model is a transformation of the ADL model. Restrictions can also be placed on the error correction model to specify other models.

From the perspective of the general-to-specific approach, potential models are viewed as restricted versions of more general models. As the name of the approach suggests, model selection proceeds by estimating a more general model, maybe more general than is theoretically suggested, and testing whether restrictions can be placed on that model. This is the procedure we followed when testing whether an LDV, rather than an ADL(1,1), model suffices to model U.S. public responsiveness.

Continuing with the general-to-specific principle, we saw that the static model with serially correlated errors was a restricted form of the ADL(1,1) model. A reasonable question at this point is "How does one choose whether to address the problem of serially correlated errors with a more general LDV or ADL model or a more specific static model with autoregressive errors or serial correlation robust standard errors?" The decision to use one or the other depends on a number of criteria.

1. If we believe that the serial correlation is a nuisance produced by some problem, such as measurement error, we should use a static model and either include autoregressive errors or estimate serial correlation robust standard errors. The problem of measurement error resulting in serial correlation can occur when the measurement error in one period depends on the measurement error in another, such as when measurement errors accumulate over time.

2. If we believe that the serial correlation is a product of a very real dynamic in the data-generating process (e.g., vote intention today is a function of vote intention yesterday), we should model this dynamic with an LDV model.

The second criterion is related to the issue of exogeneity we discussed in Chapter 3. Recall the following example data model of vote intention:

$$\text{vote}_t = \beta_0 + \beta_1 \text{econ}_t + \varepsilon_t. \qquad (4.3.1)$$

It is highly probable that the data-generating process for vote intention includes past intention, vote$_{t-1}$. If this is the case, then ε_t in our data model will be a function of (contain) vote$_{t-1}$ (i.e., vote$_{t-1}$ is omitted). If econ$_t$ are subjective evaluations of the economy, it is unlikely that vote$_{t-1}$ and econ$_t$ are independent. An individual's past vote intention is likely to be predictive of current economic evaluations (Evans & Andersen, 2006), and so it is not likely that we have contemporaneous exogeniety: $E\left(\varepsilon_t \mid econ_t\right) \neq 0$. This may be resolved by explicitly including vote$_{t-1}$ in our data model as follows:

$$\text{vote}_t = \beta_0 + \beta_1 \text{econ}_t + \beta_2 \text{vote}_{t-1} + \varepsilon_t. \tag{4.3.2}$$

In this data ε_t model, the values are not a function of vote$_{t-1}$. Including a lagged dependent variable can correct violations of endogeneity, while running a static model with AR(1) errors does not.

Continuing our discussion of when to choose a dynamic model with a lagged dependent variable or a static model with a correction for serial correlation, consider our LDV model of U.S. public policy responsiveness (Table 4.3). What if this LDV model exhibited serially correlated errors? This raises an important issue. Serial correlation when including a lagged dependent variable has serious consequences. In the context of our example, the LDV process with serially correlated errors looks as follows:

$$R_t = \alpha_0 + \alpha_1 R_{t-1} + \beta_1 P_t + \beta_2 W_t + \mu_t, \tag{4.3.3}$$

$$\mu_t = \rho \mu_{t-1} + \varepsilon_t. \tag{4.3.4}$$

From Equation 4.3.3,

$$R_{t-1} = \alpha_0 + \alpha_1 R_{t-2} + \beta_1 P_{t-1} + \beta_2 W_{t-1} + \mu_{t-1}. \tag{4.3.5}$$

Note: R_{t-1} and μ_t are both functions of μ_{t-1}. Therefore, $E\left(\mu_t \mid R_{t-1}\right) \neq 0$, and we have not met the assumptions necessary for an unbiased OLS estimation. Note that this isn't just a violation of strict exogeneity. It is a violation of contemporaneous exogeneity. Therefore, we do not even have the conditions for an asymptotically unbiased OLS estimation.

The presence of serial correlation in such an LDV or ADL model is seen as evidence of model misspecification, which can be corrected by including additional lags of the independent or dependent variables. For example, consider a single realization of the following data-generating process:

$$y_t = \alpha_0 + \alpha_1 y_{t-1} + \alpha_2 y_{t-2} + \beta_1 x_t + \varepsilon_t, \tag{4.3.6}$$

with $\alpha_0 = 10$, $\alpha_1 = 0.4$, $\alpha_2 = 0.3$, $\beta_1 = 0.9$, and $\varepsilon_t \sim N(0, 1)$. In addition to including a lag of y_t as an independent variable, it also includes the second lag. Let us use a single realization, with 38 time points, to estimate a model without the second lag—an LDV model (Table 4.4).

If we test the residuals from this model, we will find that they are not a white noise process. The Q statistic is 33.29, and the corresponding P value is 0.010. The rejection of the null hypothesis of a white noise process is because the errors contain serial correlation due to the omission of the second lag of y_t. If we reestimate this model, including the second lag of y_t, we would find that we cannot reject the null hypothesis that the residuals are a white noise process. The Q statistic is 19.48, and the corresponding P value is 0.245. It is only by including the second lag of y_t that we meet one of the necessary conditions for an asymptotically unbiased OLS estimation.

The requirement for consistent OLS estimation with a lagged dependent variable is sometimes stated as *dynamic* completeness (Wooldridge, 1991, 2006, pp. 400–402). This simply means that there are enough lags of the dependent and independent variables in the model to ensure that $E\left(\mu_t \mid y_{t-1+h}, x_{t+h}\right) = 0$, for all $h \leq 0$. Recall that this is called *sequential exogeneity.* Dynamic completeness also implies that there is no serial correlation.

Some simulation work (Keele & Kelly, 2006) has shown that under some circumstances the bias from serial correlation with a lagged dependent variable is relatively minor and typically less than the bias caused by estimating a model without a lagged dependent variable when the data-generating process for the observed time series contains a lagged dependent variable. While this certainly suggests that the analyst should not shy away from models with lagged dependent variables, it is not a justification for ignoring dynamic completeness. It is relatively easy to detect residual serial correlation using the techniques discussed in Chapter 3. Such tests should always be conducted, and if serial correlation is detected, the model

Table 4.4 LDV Model With AR(2) Data

y_t	Coefficient	Standard Error	t Statistic	P Value
y_{t-1}	0.50	0.13	3.81	0.001
x_t	0.79	0.17	4.50	<0.001
Constant	17.57	4.75	3.70	0.001

NOTE: $R^2 = 0.45$, $T = 38$; T = number of time points, AR = autoregressive, LDV = lagged dependent variable.

should be respecified with additional lags of the independent and/or the dependent variable to remove the serial correlation.[3]

An additional consideration when choosing between a dynamic and a static model is the following:

3. The dynamics for the effects of the independent variables is different between the two types of models. An LDV model has different short-run and long-run effects. For the static model with autoregressive errors (or not), the long-run effect is the short-run effect.

However, one can always add dynamics to the static model, allowing for different short-run and long-run effects, by including lags of the independent variables, as we did in the FDL model discussed in Chapter 3. Autoregressive errors can be added to an FDL model without any problem.

A fourth consideration is as follows:

4. A lagged dependent variable can *never* be strictly exogenous, as was demonstrated in Section 4.1. As strict exogeneity is a requirement for unbiasedness when we do not have enough time points to rely on the assumptions for asymptotic unbiasedness, an LDV model requires an adequate number of time points.

The problem of including a lagged dependent variable in a data model with only a few time points of data for its estimation was first discussed by Hurwicz (1950), and the resulting bias is sometimes called Hurwicz bias.

As in any modelling exercise, specification of the model must take into account these important primary considerations:

1. Is the model motivated/justified by the theoretical DGP?

2. Does it test the hypotheses in which we are interested?
    o Will it tell us what we want to know?

3. Are the assumptions of the estimation technique met?

Chapter 3 provides an overview of the assumptions we have made in the estimation of the models we have considered so far. These assumptions include homoskedasticity. We now turn our attention to another type of time series model. This is a model that places the focus on the issue of heteroskedasticity.

---

[3] As we will discuss in Chapter 5, the serial correlation may also be due to an unmodelled moving average process.

4.4 Autoregressive Conditional Heteroskedasticity (ARCH) Models

For each model we have discussed, we have made the assumption of homoskedastic errors. Consider our LDV model:

$$y_t = \alpha_0 + \alpha_1 y_{t-1} + \beta_1 x_t + \varepsilon_t. \tag{4.4.1}$$

Recall that ε_t is a white noise process. We assume that ε_t has a zero mean and a constant variance, σ^2. We might want to consider the possibility that the variance changes over time. As noted in Chapter 3, this has implications for the estimation of the standard errors of our model parameters and for subsequent inferential tests. The dynamics within the variance of the errors may also be of substantive interest—for example, a reduction in the variance in support for a U.S. president in response to critical events, such as economic crises and foreign conflict (Gronke & Brehm, 2002). Therefore, we may be interested in modelling the dynamics in the variance of ε_t. To do this, we must make the distinction between the unconditional variance of ε_t and the conditional variance of ε_t.

Considering again our LDV model, we assume that the unconditional variance of ε_t is constant and without serial correlation:

$$E\left(\varepsilon_t \varepsilon_{t-s}\right) = \begin{cases} \sigma^2 \text{ for } s = 0 \\ 0 \text{ otherwise} \end{cases}. \tag{4.4.2}$$

Note that the expected value of the squared errors is the variance of ε_t. Continuing to assume constant unconditional variance, we allow that the variance conditional on past values of the errors will change over time. Until now, we have assumed that the conditional variance of the errors is constant. Now we model the errors as a process with a conditional variance, $E(\varepsilon_t \varepsilon_t | \varepsilon_{t-s})$, that is a function of past values of the variance. A common way to do this is to model the squared errors as an autoregressive AR(m) process:

$$\varepsilon_t^2 = \zeta + \phi_1 \varepsilon_{t-1}^2 + \phi_2 \varepsilon_{t-2}^2 + \cdots + \phi_m \varepsilon_{t-m}^2 + \omega_t, \tag{4.4.3}$$

where ω_t is also a white noise process with a zero mean and constant unconditional variance. We assume that this process is covariance stationary and that $\phi_i \geq 0$, $i = 1, \ldots, m$. As ε_t^2 cannot be negative, we also assume that $\zeta > 0$ and $\omega_t \geq -\zeta$ for all t. This is called an autoregressive conditional heteroskedastic process:

$$\varepsilon_t^2 \sim \text{ARCH}(m), \qquad (4.4.4)$$

with m denoting the order of the ARCH process (Engle, 1982). Time series models that include such a process in the modelling of the errors are called ARCH models. The logic of this model of the squared errors is that periods of high (low) variance tend to group together. If the variance is higher (lower) than average at a particular time point, it is also likely to be higher (lower) than average in the next time point. Such ARCH errors can be included in static, autoregressive, or ARMA (to be discussed in the next chapter) models.

An equivalent, and common, representation of an ARCH process is as follows:

$$\varepsilon_t = \sqrt{h_t} \, v_t, \qquad (4.4.5)$$

where v_t is a white noise process with a zero mean and unit variance and h_t is a function of past ε_t^2 (Hamilton, 1994). Specifically,

$$h_t = \zeta + \phi_1 \varepsilon_{t-1}^2 + \phi_2 \varepsilon_{t-2}^2 + \cdots + \phi_m \varepsilon_{t-m}^2.$$

$$E(v_t) = 0$$

$$E(v_t v_{t-s}) = \begin{cases} 1 \text{ for } s = 0 \\ 0 \text{ otherwise} \end{cases}. \qquad (4.4.6)$$

We can see that in this representation, the unconditional expected value of ε_t is 0, as assumed in Equation 4.4.1:

$$E(\varepsilon_t) = E\left(\sqrt{h_t} \, v_t\right) = E\left(\sqrt{h_t}\right) E(v_t) = 0. \qquad (4.4.7)$$

Note that the second equality utilizes the assumption that v_t is a white noise process. The conditional expected value is also equal to 0 (Enders, 2004).

$$E(\varepsilon_t \mid \varepsilon_{t-1}, \varepsilon_{t-2}, \cdots, \varepsilon_{t-m}) = 0. \qquad (4.4.8)$$

The unconditional variance of ε_t is a constant, which we have defined as σ^2 in Equation 4.4.2. In particular,

$$E\left(\varepsilon_t^2\right) = E\left(h_t v_t^2\right) = E\left(h_t\right),$$

$$= E\left(\zeta + \phi_1 \varepsilon_{t-1}^2 + \phi_2 \varepsilon_{t-2}^2 + \cdots + \phi_m \varepsilon_{t-m}^2\right),$$

$$= \frac{\zeta}{1 - \phi_1 - \phi_2 - \cdots - \phi_m}. \tag{4.4.9}$$

Finally, the conditional variance of ε_t, given past ε_t, is a function of time. Specifically,

$$E\left(\varepsilon_t^2 \mid \varepsilon_{t-1}, \ldots, \varepsilon_{t-m}\right) = E\left(h_t \mid \varepsilon_{t-1}, \ldots, \varepsilon_{t-m}\right) E\left(v_t^2\right),$$

$$= h_t = \zeta + \phi_1 \varepsilon_{t-1}^2 + \phi_2 \varepsilon_{t-2}^2 + \cdots + \phi_m \varepsilon_{t-m}^2. \tag{4.4.10}$$

Note that the conditional variance is an autoregressive process but the errors themselves (ε_t) are not.

We return to our example of economic popularity in West Germany, but now for a longer time period, 1977–1998, and in place of vote intention data, we use government approval data based on the following survey question: "Sind Sie mit dem was die jetzige CDU/CSU/FDP-Regierung in Bonn bisher geleistet hat eher zufrieden oder eher unzufrieden?" (Are you rather satisfied [or happy] or rather dissatisfied [or unhappy] with the performance of the current CDU/CSU/FDP government in Bonn thus far?)

During this period, we expect periodicity from one electoral cycle to the next, separate trends for the period in which the SPD (with the FDP) controlled government and for the period in which the CDU (with the FDP) controlled government, and heteroskedasticity. The latter is expected due to changes in the party system prompted by the reunification of East and West Germany. During this time, partisan identification transitioned from being quite strong for many West Germans to quite weak (Arzheimer, 2006; Pickup, 2010; Zelle, 1998).

We begin by partialling out the periodicity and trending in the government approval data. We regress government opposition on two periodicity terms and a separate trend for each of the governments.[4]

---

[4] The two periodicity variables allow us to model a flexible election cycle, where not only is the amplitude of the cycle estimated but so is the timing of the peaks and troughs (phase).

The results are presented in Table 4.5. Estimated residuals from this regression are the approval data with periodicity and trending partialled out. Using this data, we estimate an ADL(1,1) economic popularity model. The independent variables included are GDP growth, unemployment, and inflation. The results from this estimation are presented in Table 4.6.

None of the coefficients on the economic variables are statistically significant on their own. We can test the joint significance of each economic variable with its lag using an F statistic. For GDP growth, the $F(1, 234)$ statistic is 2.12, and the corresponding P value is 0.147. For inflation the $F(1, 234)$ statistic is 6.33, and the corresponding P value is 0.013. For unemployment, the $F(1, 234)$ statistic is 8.38, and the corresponding P value is 0.004. We can also calculate the long-run effects for inflation and unemployment and test their significance. For inflation, the long-run effect is $(0.32-0.79)/(1-0.78) = -2.14$. The chi-squared(1) statistic is 6.81 with a P value of 0.010. For unemployment, the long-run effect is $(0.23-0.77)/(1-0.78) = -2.45$. The chi-squared(1) statistic is 10.95 with a P value of 0.001. Inflation and unemployment have statistically significant long-run effects.[5]

Table 4.5　West Germany Government Approval 1977–1998

Approval	Coefficient	Standard Error	t Statistic	P Value
SPD trend	0.035	0.031	1.13	0.258
CDU trend	−0.13	0.0095	−13.83	<0.001
Cycle 1	5.83	0.71	8.26	<0.001
Cycle 2	0.24	0.71	0.34	0.736
Constant	61.047	1.04	58.98	<0.001

NOTE: $R^2 = 0.59$, $T = 243$; T = number of time points.

The cycling is captured by including two terms in the regression: $\beta_1 \sin(\lambda\theta)$ and $\beta_2 \cos(\lambda\theta)$, where β_1 and β_2 are the parameters to be estimated and λ is the frequency of the election cycle, so that $\lambda\theta$ is defined by the length of the interelection period. If you plot the sine and cosine functions, they represent waves. As $\lambda\theta$ increases over a range of 2π, both the sine and the cosine functions cycle through a complete peak and trough. The sine and cosine functions are out of phase with each other, in that when one is at its maximum the other is halfway between its maximum and minimum. By combining a sine and a cosine function, weighted by β_1 and β_2, respectively, we can create a wave of any amplitude (and phase). The estimated parameters β_1 and β_2 can be used to calculate the amplitude of the interelection cycle.

[5] If we tested the long-run effect of GDP, we would find it is not statistically significant at the 0.05 significance level.

Table 4.6   West Germany Government Approval 1977–1998, ADL(1,1)

Approval	Coefficient	Standard Error	t Statistic	P Value
L1. Approval	0.78	0.041	18.80	<0.001
Growth	0.41	0.23	1.75	0.081
Unemployment	0.23	1.85	0.13	0.900
Inflation	0.32	0.87	0.37	0.715
L1. Growth	−0.27	0.24	−1.13	0.259
L1. Unemployment	−0.77	1.88	−0.41	0.684
L1. Inflation	−0.79	0.85	−0.93	0.352
Constant	4.27	1.54	2.77	0.006

NOTE: $R^2 = 0.76$, $T = 242$; ADL = autoregressive distributed lag, T = number of time points, L1 = first lag.

Before settling on these results, we would want to test the estimated errors of the model for, among other things, heteroskedasticity. We do this here with Engle's Lagrange multiplier test for the presence of autoregressive conditional heteroscedasticity (Engle, 1982). Specifically, we test the null of no ARCH effects against the alternative of ARCH(m) effects. This test is conducted on the residuals of the model we just ran: $\hat{\varepsilon}_t$. We run the following auxiliary regression: $\hat{\varepsilon}_t^2 = \gamma_0 + \gamma_1 \hat{\varepsilon}_{t-1}^2 + \cdots + \gamma_p \hat{\varepsilon}_{t-m}^2$, for up to m lags of the squared residuals and test the null hypothesis that $\gamma_1 = \cdots = \gamma_m = 0$ using a Lagrange multiplier test (Greene, 2003, pp. 484–492). We select m on the basis of the order of the ARCH effects for which we wish to test. In our current example, we will test for an ARCH(1) process.

This test is applicable for both the alternatives of autoregressive and moving average processes within the variances. We will discuss the latter type of process in Chapter 5. For our example, we test for first-order ARCH effects. The resulting test statistic is 36.21 with a chi-squared distribution with 1 degree of freedom. The corresponding P value is <0.001. This means that we can reject the null hypothesis of no first-order ARCH effects. Based on this finding, we now estimate the West German economic approval model as an ADL(1,1) with an ARCH(1) term (Table 4.7). As with the static model with

Table 4.7 West Germany Government Approval 1977–1998, ADL(1,1) ARCH(1)

Approval	Coefficient	Standard Error	z Statistic	P Value
L1. Approval	0.84	0.044	19.17	<0.001
Growth	0.50	0.31	1.64	0.100
Unemployment	−0.11	3.62	−0.03	0.976
Inflation	0.036	0.93	0.04	0.969
L1. Growth	−0.41	0.29	−1.41	0.160
L1. Unemployment	−0.31	3.62	−0.09	0.931
L1. Inflation	−0.48	0.92	−0.52	0.602
Constant	3.65	1.44	2.53	0.011
ARCH				
L1.	0.21	0.061	3.37	0.001
Constant	11.29	0.70	16.04	<0.001

NOTE: Log likelihood = −658.07, $T = 242$; L1 = first lag, T = number of time points, ADL = autoregressive distributed lag, ARCH = autoregressive conditional heteroskedasticity.

autoregressive errors, ARCH models are typically estimated using maximum likelihood.

First, we note that the ARCH(1) term is statistically significant. This suggests that the inclusion of the ARCH term is appropriate. As an alternative, we could have just estimated the ADL(1,1) and estimated heteroskedasticity robust standard errors. However, OLS is inefficient in the presence of heteroskedasticity, so there are efficiency gains (smaller standard errors) to be made by including the ARCH term. Furthermore, the statistical significance of the ARCH term may be of substantive interest.

As we have done previously, we can test the significance of the short-run effects using the coefficient on the contemporaneous values of each economic variable. In each case the P value is greater than 0.05 and we cannot reject the null hypothesis of no short-run effect. As for the estimated long-run effects for inflation and unemployment, these are as follows.

For inflation, the long-run effect is $(0.036-0.48)/(1-0.84) = -13.38$. The chi-squared(1) statistic is 4.19 with a P value of 0.041. For unemployment, the long-run effect is $(-0.11-0.32)/(1-0.84) = -15.31$ The chi-squared(1) statistic is 5.58 with a P value of 0.018. Again, a change in inflation or unemployment has a statistically significant effect on approval.

We can also model the variance of ε_t as a function of exogenous variables. This is called multiplicative heteroskedasticity. For example, we can have an ARCH(1) process in which the variance is also a function of the variable x_t:

$$h_t = exp\left(\zeta_0 + \zeta_0 x_t\right) + \phi_1 \varepsilon_{t-1}^2.$$

When we discuss generalized autoregressive conditional heteroskedasticity (GARCH) models in Chapter 5, we will discuss how to test the adequacy of the specification of ARCH and GARCH models. We now turn to yet another type of time series model. This model assumes that the conditional variance is constant but allows for serial correlation in the errors of a form that differs from autoregressive errors.

4.5 The Moving Average Process and the Autocorrelated Error Model

A moving average data-generating process can be characterized as

$$y_t = \beta_0 + \varepsilon_t + \sum_{j=1}^{q} \phi_j \varepsilon_{t-j}, \tag{4.5.1}$$

where q determines the order of the moving average process, the ϕ_j values are coefficients with at least one $\phi_j \neq 0$, and $\varepsilon_t \sim NID\left(0, \sigma_\varepsilon^2\right)$. Note that a moving average process differs from an autoregressive process in that the first includes lags of the errors while the second includes lags of the dependent variable. It also differs from a static process with autoregressive errors in that the moving average process includes lags of the errors in the equation for y_t, while the static model includes lags of the errors in the equation for the errors. As an example, a moving average process of order 1 includes one lag of the error:

$$y_t = \beta_0 + \varepsilon_t + \phi_1 \varepsilon_{t-1}. \tag{4.5.2}$$

Such a process is denoted as MA(1). This is a weakly dependent sequence as y_t values one period apart are correlated but y_t values two periods apart are not. This can be demonstrated as follows: $y_t = \varepsilon_t + \phi_1 \varepsilon_{t-1}$ and $y_{t-1} = \varepsilon_{t-1} + \phi_1 \varepsilon_{t-2}$ have ε_{t-1} in common and are therefore correlated, but $y_t = \varepsilon_t + \phi_1 \varepsilon_{t-1}$ and $y_{t-2} = \varepsilon_{t-2} + \phi_1 \varepsilon_{t-3}$ have nothing in common and are therefore not correlated.

A moving average process can be a data model in itself. It is called the autocorrelated error model or just an MA model. Exogenous regressors (x_k) can also be included in such data models:

$$y_t = \beta_0 + \sum_{k=1}^{k} \beta_k x_k + \varepsilon_t + \sum_{j=1}^{q} \phi_j \varepsilon_{t-j}. \qquad (4.5.3)$$

Like ARCH models and static models with autoregressive errors, MA models are estimated using maximum likelihood. If we were to estimate Equation 4.5.3 without the moving average component, we would have serial correlation in the errors. A moving average process is another source of violation of the assumption of no serial correlation. Therefore, an MA model can be useful to control for the moving average component of a data-generating process. As an example of an MA model, consider the following data on U.S. Environmental Protection Agency (EPA) enforcements.[6] To begin, we look at a plot of the number of enforcements over time (Figure 4.1).

The data are monthly, and the first time point $(t = 1)$ is January 2001. The last data point is December 2005. These data were used by Provost, Gerber, and Pickup (2009) to test the impact of the various tools President Bush had at his disposal to influence EPA environmental regulation. They considered the impact of budgetary changes, the appointment and resignation of various EPA administrators, and the introduction of two major rule changes. The first major change was to the New Source Review (NSR) rules, and the second was the issuance of the Clean Air Interstate Rule (CAIR).[7]

With "enforcements" as the dependent variable, we estimate the following data model with exogenous regressors (independent variables) and a single second-order moving average term:

$$y_t = \beta_0 + \sum_{k=1}^{K} \beta_k x_k + \varepsilon_t + \phi_j \varepsilon_{t-2}. \qquad (4.5.4)$$

---

[6] Enforcement activities consist of notices of violation, consent decrees, and administrative orders.

[7] Each is coded as "1" for the month after which the change occurred and for the 2 months following that; otherwise, they are coded as "0."

104

Figure 4.1 U.S. Environmental Protection Agency Enforcements, 2001 to 2005

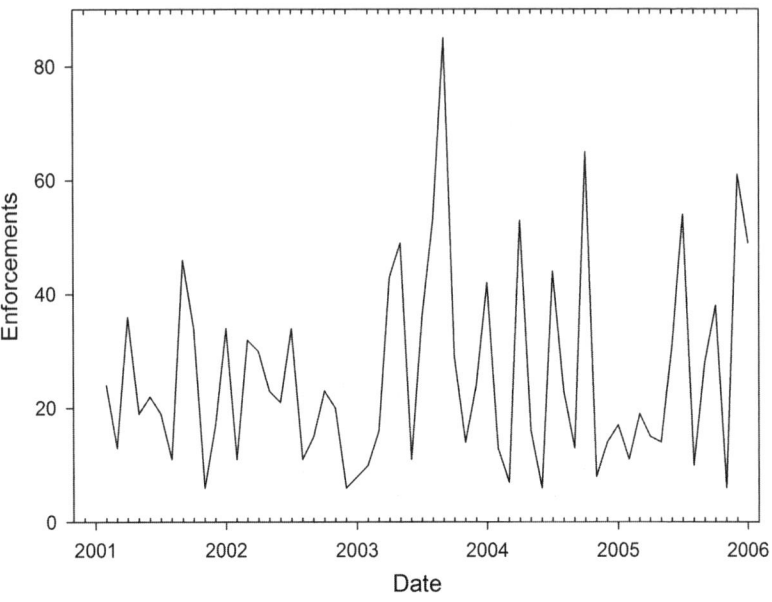

The *K* regressors included in this model are (a) increases in the budget and decreases in the budget—increases and decreases are measured in millions of U.S. dollars and separated, as they are expected to have different effects; (b) the appointments of Whitman, Holmstead, Leavitt, Johnson, and Sansonetti as EPA administrators; (c) Whitman's resignation; (d) the introduction of the NSR; and (e) the introduction of the CAIR. Note that we have not included a first-order moving average term. This is a perfectly acceptable modelling choice, but it is fair to ask at this point how it was decided to use a second-order MA model for these data. This topic is covered in Chapter 5, along with that of determining the number of lags of the dependent variable to include in a model. The results of the maximum likelihood estimation of the model are presented in Table 4.8.

Certain policy initiatives of the Bush administration did have a clear impact on EPA enforcements. A million-dollar decrease in the budget resulted, on average, in 86 less enforcements. Budget increases, on the other hand, do not seem to have had any effect. Changes to the NSR rules did have a reducing effect on enforcements, which arguably was the intended effect (Provost et al., 2009).

Table 4.8   MA Model of U.S. Environmental Protection Agency
Enforcements, 2001 to 2005

Enforcements	Coefficient	Standard Error	t Statistic	P Value
Budget increase	−8.84	27.15	−0.33	0.745
Budget decrease	86.45	29.19	2.96	0.003
Whitman	−30.64	6.06	−5.05	<0.001
Whitman quit	9.65	17.17	0.56	0.574
Holmstead	1.00	11.85	0.08	0.932
Leavitt	−22.29	5.53	−4.03	<0.001
Johnson	−15.40	25.09	−0.61	0.539
Sansonetti	−5.08	8.95	−0.57	0.571
NSR	−17.68	8.95	−1.97	0.048
CAIR	5.80	17.92	0.32	0.746
Constant	56.15	10.40	5.40	<0.001
L2. MA	−0.66	0.12	−5.45	<0.001

NOTE: Log likelihood = −235.22, T = 59; T = number of time points, L2 = second lag,
CAIR = Clean Air Interstate Rule, MA = moving average, NSR = New Source Review.

It is also clear that the coefficient for the second-order moving average process is statistically significant at the usual 0.05 significance level. Interestingly, the coefficient is negative. It is generally difficult to give a substantive interpretation of moving average processes. In this particular instance, it is probably best to just treat it as a nuisance that needs to be dealt with in order to get the correctly estimated values for the other parameters (and their standard errors) in the model.

We can now use the Q test to test the residuals from the EPA enforcement model for deviation from a white noise process. The Q statistic is 31.21. It is chi-squared distributed with 27 degrees of freedom, giving us a P value of 0.26. Based on this test, we cannot reject the null hypothesis of a white noise process at the 0.05 significance level. Therefore, there is no evidence that any serial correlation remains within the residuals. For illustration, we can also run the same model, but without the moving average component, and test the residuals for deviation from a white noise process. The results are presented in Table 4.9.

Table 4.9 Model of U.S. Environmental Protection Agency
Enforcements, 2001 to 2005

Enforcements	Coefficient	Standard Error	t Statistic	P Value
Budget increase	22.53	51.66	0.44	0.663
Budget decrease	75.63	53.48	1.41	0.157
Whitman	−32.38	8.22	−3.94	<0.001
Whitman quit	−11.27	19.62	−0.57	0.566
Holmstead	2.75	15.82	0.17	0.862
Leavitt	−22.61	8.19	−2.76	0.006
Johnson	−18.13	25.75	−0.70	0.481
Sansonetti	−1.96	12.12	−0.16	0.871
NSR	−24.54	10.29	−2.39	0.017
CAIR	7.37	13.13	0.56	0.575
Constant	49.79	19.01	2.62	0.009

NOTE: Log likelihood = −242.71, T = 59; P = probability, T = number of time points, CAIR = Clean Air Interstate Rule, NSR = New Source Review.

The Q statistic is now 49.65, giving us a P value of 0.005. We reject the null hypothesis of a white noise process at the 0.05 significance level. This indicates that there is serial correlation in the residuals. Also, note that when we do not control for the moving average process, the effect of budget decreases is no longer statistically significant. As in this example, there is often no intuitive interpretation for a moving average process or model, but it is important to include it in order to control for serial correlation in the data model residuals.

Now let's consider the necessary conditions for stationarity for the time series processes we have discussed in this chapter.

4.6 Stability and Stationarity Conditions

In Chapter 2, we discussed the importance of meeting the assumption of covariance stationarity. This would be a good time to discuss the conditions that are necessary for a dynamic model to meet this assumption. We start with the stationarity conditions for an AR(1) process. We do this by revisiting the topic of stability, first discussed in Chapter 2. Stability is a necessary condition for stationarity.

Time series analysis is concerned with describing and modelling time series sequences $Y_t = \{y_1, y_2, \ldots, y_T\}$ with equations that contain both nonstochastic and stochastic components. The LDV model without independent variables is an example of such an equation. It contains ε_t (the stochastic component), one lag of y_t, and a constant:

$$y_t = \alpha_0 + \alpha_1 y_{t-1} + \varepsilon_t \cdot \qquad (4.6.1)$$

This assumes an AR(1) data-generating process with a constant α_0 (and therefore a nonzero equilibrium). Statisticians often talk of a "solution" to the equation describing a data-generating process. The solution expresses the value of y_t as a function of the elements of the $\{x_t\}$ sequence (if there is any x_t in the model), ε_t, the data-generating process parameters, and t (time). The solution does not contain lags of the dependent variable, even though the equation for the data-generating process does. The solution may possibly contain some initial conditions for the sequence $\{y_t\}$. The initial condition is the value of y_t when the time series process began. It is sometimes denoted as y_0, but it may not necessarily correspond with the first data point in our data set; we may not observe the data-generating process when it first began.

The solution allows us to do something we cannot do with the equation itself. It allows us to talk in the abstract about what values we would expect y_t to take on average; that is, we can calculate the mean value of y_t over the long run. This is the unconditional expected value.

Earlier in this chapter, we learned that the estimated long-run equilibrium of the LDV model is $\alpha_0/(1 - \alpha_1)$. This is the unconditional expected value. We will now see how this was derived from the solution to the LDV data-generating process. As the solution allows us to calculate the value we expect y_t to take (on average) over the long run, we can use it to determine whether a time series has a stable or an explosive time path. If the time series is stable, the unconditional expected value (long-term mean) will remain constant over the long run—not change over time. We cannot determine this with an equation containing a lag of y_t. Why not?

Take, for example, $y_t = \alpha_0 + \alpha_1 y_{t-1} + \varepsilon_t$, and say we know α_0 and α_1 (which we normally would not). We know the expected value of $\{\varepsilon_t\}$: $E(\varepsilon_t) = 0$; but to calculate the expected value of y_t, we need to know the expected value of y_{t-1}. But this is the same as knowing the expected value of y_t, which is what we are trying to determine; that is, we do not know the expected value of y_{t-1} any more than we know the expected value of y_t. We can plug $y_{t-1} = \alpha_0 + \alpha_1 y_{t-2} + \varepsilon_{t-1}$ into $y_t = \alpha_0 + \alpha_1 y_{t-1} + \varepsilon_t$ and get

$$y_t = \alpha_0 + \alpha_1 \left(\alpha_0 + \alpha_1 y_{t-2} + \varepsilon_{t-1} \right) + \varepsilon_t,$$

$$= \alpha_0 + \alpha_0 \alpha_1 + \alpha_1 \alpha_1 y_{t-2} + \varepsilon_t + \alpha_1 \varepsilon_{t-1}.$$

But what is the expected value of y_{t-2}? Well, $y_{t-2} = \alpha_0 + \alpha_1 y_{t-3} + \varepsilon_{t-2}$, and we can plug this in, but we are going to need to plug in y_{t-3}, and so on. If we repeated this $t + m$ times, we would end up with the following:

$$y_t = \alpha_0 \sum_{i=0}^{t+m} \alpha_1^i + \alpha_1^{t+m+1} y_{-(m+1)} + \sum_{i=0}^{t+m} \alpha_1^i \varepsilon_{t-i} \text{ for all } t > 0. \quad (4.6.2)$$

$t + m$ is an arbitrarily large number of time points backward in time. It is as many times as we may wish to replace y_{t-j} with $y_{t-j} = \alpha_0 + \alpha_1 y_{t-j-1} + \varepsilon_{t-j}$. There is a lot of mathematical notation involved here, but it can be simplified. If $|\alpha_1| < 1$, the term α_1^{t+m+1} approaches 0 as m approaches infinity, and the infinite sum $\sum_{i=0}^{t+m} \alpha_1^i = 1 + \alpha_1 + \alpha_1^2 + \cdots)$ converges to $1/(1 - \alpha_1)$. Therefore, as long as $|\alpha_1| < 1$ and the time series began a long time ago—allowing us to replace y_{t-j} many, many times—we can rewrite Equation 4.6.2 as

$$y_t = \frac{\alpha_0}{(1 - \alpha_1)} + \sum_{i=0}^{\infty} \alpha_1^i \varepsilon_{t-i}. \quad (4.6.3)$$

This is a solution to the original equation: We have expressed y_t as a function of the parameters of the data-generating process, ε_t, and time. Since $E(\varepsilon_t) = 0$, the unconditional expected value of y_t is

$$E(y_t) = \frac{\alpha_0}{(1 - \alpha_1)}. \quad (4.6.4)$$

As we noted before, $\alpha_0/(1 - \alpha_1)$ is the equilibrium to which the process will converge in the absence of external shocks. Hence, we use $\hat{\alpha}_0/(1 - \hat{\alpha}_1)$ to estimate the long-run equilibrium. As the process does have a constant value to which it will converge, it is stable.

If our lagged dependent variable process contains an exogenous variable, $y_t = \alpha_0 + \alpha_1 y_{t-1} + \beta_1 x_t + \varepsilon_t$, Equation 4.6.2 would be

$$y_t = \alpha_0 \sum_{i=0}^{t+m} \alpha_1^i + \alpha_1^{t+m+1} y_{-(m+1)} + \sum_{i=0}^{t+m} \alpha_1^i \beta_1 x_{t-i}$$

$$+ \sum_{i=0}^{t+m} \alpha_1^i \varepsilon_{t-i} \text{ for all } t > 0. \tag{4.6.5}$$

If we assume that x_t has been constant since before $t + m$ time periods (a very long time ago), $x_t = c$, then Equation 4.6.3 would be

$$y_t = \frac{\alpha_0}{(1-\alpha_1)} + \frac{\beta_1}{(1-\alpha_1)} c + \sum_{i=0}^{\infty} \alpha_1^i \varepsilon_{t-i}. \tag{4.6.6}$$

If we instead assume that x_t increased by one unit $t + m$ time periods ago and has not changed since, $x_t = c + 1$, then Equation 4.6.3 would be

$$y_t = \frac{\alpha_0}{(1-\alpha_1)} + \frac{\beta_1}{(1-\alpha_1)}(c+1) + \sum_{i=0}^{\infty} \alpha_1^i \varepsilon_{t-i}. \tag{4.6.7}$$

The difference between Equations 4.6.6 and 4.6.7 is $\beta_1/(1-\alpha_1)$. Hence, we use $\hat{\beta}_1/(1 - \hat{\alpha}_1)$ to estimate the long-run effect of a one-unit change in x_t, as we saw earlier.

Returning to the issue of stability, if $\alpha_1 = 1$, the term $\sum_{i=0}^{t+m} \alpha_1^i = 1 + \alpha_1 + \alpha_1^2 + \cdots$ in Equation 4.6.2 does not converge, and the equilibrium is undefined: $E(y_t) = ?$ If $|\alpha_1| > 1$, the term α_1^{t+m+1} in Equation 4.6.2 will tend to infinity over time: $E(y_t) = \pm\infty$.

Therefore, the data-generating process $y_t = \alpha_0 + \alpha_1 y_{t-1} + \varepsilon_t$ is stable if $|\alpha_1| < 1$, but not otherwise. As stability is a necessary condition for stationarity, the data-generating process $y_t = \alpha_0 + \alpha_1 y_{t-1} + \varepsilon_t$ cannot be stationary unless $|\alpha_1| < 1$.

This is true for any first-order autoregressive process: An AR(1) process is stable if $|\alpha_1| < 1$. In addition to stability, stationarity requires that the data-generating process contain no trending, periodicity, or structural breaks. In addition to this, we require one further condition for stationarity. When a time series process begins, it may not begin in equilibrium. In the convergent examples we looked at in Chapter 2, they each started at 1 and eventually converged on the equilibrium of 0. If we happen to be measuring this time series during this period of equilibration, our observed time series will not be stationary (the mean, variance, and covariances will be changing during this period), even though the series is stable in the long run. Therefore, we need to assume that the sequence started long enough ago so that it has had a chance to equilibrate, or that the process started immediately in equilibrium.

For the AR(1) process, note how important the value of α_1 is to stability. For the AR(1) equation, α_1 is called the characteristic root (denoted as m_1). For an AR(1) process to be stable, the absolute value of its characteristic root must be less than 1.

We now consider higher-order autoregressive processes. A second-order autoregressive process is written as

$$y_t = \alpha_0 + \alpha_1 y_{t-1} + \alpha_2 y_{t-2} + \varepsilon_t. \tag{4.6.8}$$

The stability conditions for such an AR(2) process are as follows (Harvey, 1993, pp. 18–19):

$$\alpha_1 + \alpha_2 < 1.$$

$$-\alpha_1 + \alpha_2 < 1.$$

$$\alpha_2 > -1. \tag{4.8.9}$$

For higher-order autoregressive processes,

$$y_t + \sum_{i=1}^{p} \alpha_i y_{t-i} + \varepsilon_t. \tag{4.6.10}$$

The characteristic roots ($m_1, m_2, \dots$) are a function of the parameters on the lagged dependent variables ($\alpha_1, \alpha_2, \dots$). The condition for stationarity is that the moduli (absolute value of a complex number) of the characteristic roots are less than 1. We do not need to worry about the exact functions. For higher-order autoregressive processes the characteristic roots are difficult to calculate and may take on imaginary values. Fortunately, social scientists rarely deal with anything higher than a second- or third-order autoregressive process. When we do, software packages will usually calculate the moduli of the characteristic roots for us, so that we can test stability.

There are some general rules regarding the equation parameters (α_i) that can be followed:[8]

1. In the equation describing a pth-order autoregressive process, a necessary condition for stability is $\sum_{i=1}^{p} \alpha_i < 1$.

2. A sufficient condition for stability is $\sum_{i=1}^{p} |\alpha_i| < 1$.

3. At least one characteristic root equals unity if $\sum_{i=1}^{p} \alpha_i = 1$.

---

[8] Enders (2004).

Any sequence that contains one or more characteristic roots that equal unity is called a unit root process. Such a process is not stable. For example, $y_t = \alpha_1 y_{t-1} + \varepsilon_t$, with $\alpha_1 = 1$, is a unit root process. It is often called a random walk and has very interesting properties, which we shall discuss further in Chapter 6.

For the stationarity conditions for ARCH processes, we can apply the same stationarity conditions listed above for autoregressive processes to the squared errors in the data-generating process of the ARCH model (Equation 4.4.3). For a moving average process to be stationary, it simply needs to be of a finite order ($q \neq \infty$). This is not an overly restrictive condition.

Summary

You have now been introduced to both static and dynamic models and the idea of selecting between these models using the general-to-specific approach, as well as a set of criteria to choose between using an LDV model and a static model that corrects for serial correlation in the errors.

In the next chapter, you will be introduced to another set of models called autoregressive moving average (ARMA) models. These lend themselves to another form of model selection. This is the Box-Jenkins approach, to which you will also be introduced in the next chapter.

CHAPTER 5: AUTOREGRESSIVE MOVING AVERAGE MODELS

In Chapters 3 and 4, we covered time series models that can be estimated using ordinary least squares (OLS). You were also introduced to models that are commonly estimated using maximum likelihood—the static model with AR(1) (autoregressive process of order 1) errors, the autoregressive conditional heteroskedasticity (ARCH) model, and the moving average (MA) model. In this chapter, we move on to the autoregressive moving average (ARMA) model and the Box-Jenkins approach to building such models. The chapter continues with a discussion of including exogenous regressors in our model for the purposes of estimating the magnitude of their effects and hypothesis testing. This includes a short discussion on transfer functions and intervention analysis. The chapter concludes with a discussion of an extension to ARCH models—generalized autoregressive conditional heteroskedasticity (GARCH) models.

5.1 Autoregressive Moving Average (ARMA) Models

So far, we have examined the static, finite distributed lag (FDL), lagged dependent variable (LDV), autoregressive distributed lag (ADL), ARCH, and MA models. Let us consider yet another type of time series data-generating process that we may choose to model and identify its stationarity conditions. Combining an AR(p) process with an MA(q) process (moving average process of order q) produces an ARMA data-generating process:

$$y_t = \alpha_0 + \sum_{i=1}^{p} \alpha_i y_{t-i} + \sum_{j=0}^{q} \phi_j \varepsilon_{t-j}. \qquad (5.1.1)$$

An ARMA process with a p-order autoregressive process and a q-order moving average process is denoted as ARMA(p,q). We can summarize the necessary and sufficient conditions for the stationarity of an ARMA(p,q) process as follows:

a. The process must have started an infinitely long time ago or must have immediately begun in equilibrium.

b. The autoregressive component of the time series process is stable (e.g., for ARMA(1,q), $|\alpha_1| < 1$).

c. The process cannot contain structural breaks, trending, or periodicity.

d. q must be finite.

These are really just a combination of the necessary and sufficient conditions for the stationarity of the AR(p) and MA(q) processes.

ARMA data models are used to model the time series dynamics in data that are suspected to be generated by processes of this sort. In doing so, we control for those dynamics that violate the assumptions of exogeneity and no serial correlation. Recall from Chapter 4 how including an autoregressive component may solve problems of endogeneity. Exogenous regressors (and their lags) can also be included in ARMA data models. To see how this is done, first transform the ARMA model as follows. The ARMA(p,q) model

$$y_t = \alpha_0 + \sum_{i=1}^{p} \alpha_i y_{t-i} + \sum_{j=1}^{q} \phi_j \varepsilon_{t-j} + \varepsilon_t \qquad (5.1.2)$$

can also be written as

$$y_t = \beta_0 + \mu_t,$$

$$\mu_t = \sum_{i=1}^{p} \alpha_i \mu_{t-i} + \sum_{j=1}^{q} \phi_j \varepsilon_{t-j} + \varepsilon_t, \qquad (5.1.3)$$

where the first equation is called the structural component and the second is called the disturbance component. Exogenous regressors and their lags are included in the structural component:

$$y_t = \beta_0 + \sum_{k=1} \beta_k x_k + \mu_t,$$

$$\mu_t = \sum_{i=1}^{p} \alpha_i \mu_{t-i} + \sum_{j=1}^{q} \phi_j \varepsilon_{t-j} + \varepsilon_t. \qquad (5.1.4)$$

Sometimes ARMA models with exogenous regressors—x_ks—are called ARMAX models, but we will not be making the distinction.

We can demonstrate the equivalence of Equations 5.1.2 and 5.1.3 for the ARMA(1,1) data-generating process as follows. The representation described by Equation 5.1.2 is

$$y_t = \alpha_0 + \alpha_1 y_{t-1} + \phi_1 \varepsilon_{t-1} + \varepsilon_t. \qquad (5.1.5)$$

Let $\mu_t = y_t - \dfrac{\alpha_0}{1-\alpha_1}$; therefore,

$$y_t = \mu_t + \frac{\alpha_0}{1-\alpha_1}. \tag{5.1.6}$$

And $y_{t-1} = \mu_{t-1} + \dfrac{\alpha_0}{1-\alpha_1}$. $\tag{5.1.7}$

Insert Equations 5.1.6 and 5.1.7 into Equation 5.1.5:

$$\mu_t + \frac{\alpha_0}{1-\alpha_1} = \alpha_0 + \alpha_1 \left(\mu_{t-1} + \frac{\alpha_0}{1-\alpha_1} \right) + \phi_1 \varepsilon_{t-1} + \varepsilon_t,$$

$$\mu_t = -\frac{\alpha_0}{1-\alpha_1} + \alpha_0 + \alpha_1 \frac{\alpha_0}{1-\alpha_1} + \alpha_1 \mu_{t-1} + \phi_1 \varepsilon_{t-1} + \varepsilon_t.$$

Note that $-\dfrac{\alpha_0}{1-\alpha_1} + \alpha_0 + \alpha_1 \dfrac{\alpha_0}{1-\alpha_1} = 0$, and so

$$\mu_t = \alpha_1 \mu_{t-1} + \phi_1 \varepsilon_{t-1} + \varepsilon_t. \tag{5.1.8}$$

This is the second component of the representation described by Equation 5.1.3. Next, let $\beta_0 = \dfrac{\alpha_0}{1-\alpha_1}$; then, Equation 5.1.6 becomes

$$y_t = \beta_0 + \mu_t. \tag{5.1.9}$$

This is the first component of the representation described by Equation 5.1.3.

One of the greatest difficulties of modelling a time series process with an ARMA data model is that there are commonly many alternative ARMA models that capture the same data-generating process dynamics. There is no real way of determining which model reflects the true data-generating process. Box and Jenkins (1976), therefore, proposed a method for selecting an appropriate ARMA model given the data based on a certain set of goals.

5.2 Box-Jenkins Approach to ARMA Models

Having now introduced the ARMA model, we review the Box-Jenkins approach for specifying the model. First, we need to discuss notation.

When we introduced autoregressive processes, we defined an AR(2) process as

$$y_t = \alpha_1 y_{t-1} + \alpha_2 y_{t-2} + \varepsilon_t. \qquad (5.2.1)$$

The following data-generating process is also perfectly possible:

$$y_t = \alpha_2 y_{t-2} + \varepsilon_t. \qquad (5.2.2)$$

This process contains the second lag of the dependent variable but not the first. We need some way of distinguishing between these types of models in our notation. When it is not clear from the context, we shall distinguish these as follows:

$$y_t = \alpha_2 y_{t-2} + \varepsilon_t \quad \text{AR}(p = 2),$$

$$y_t = \alpha_1 y_{t-1} + \alpha_2 y_{t-2} + \varepsilon_t \quad \text{AR}\left(p = 1, 2\right).$$

This will apply to any possible lag except the first. Since AR($p = 1$) = AR(1), we shall just use AR(1).

Furthermore, when discussing moving average processes, we defined an MA(2) process as

$$y_t = \varepsilon_t + \phi_1 \varepsilon_{t-1} + \phi_2 \varepsilon_{t-2}. \qquad (5.2.3)$$

The following data-generating process is also perfectly possible:

$$y_t = \varepsilon_t + \phi_2 \varepsilon_{t-2}. \qquad (5.2.4)$$

This is the process assumed in our example of an MA model earlier, in Chapter 4. When it is not clear from the context, we shall distinguish these as follows:

$$y_t = \varepsilon_t + \phi_2 \varepsilon_{t-2} \text{MA}(q = 2).$$

$$y_t = \varepsilon_t + \phi_1 \varepsilon_{t-1} + \phi_2 \varepsilon_{t-2} \text{MA}(q = 1, 2).$$

Again, this will apply to any possible lag except the first. Since MA($q = 1$) = MA(1), we shall just use MA(1).

We are now in a position to discuss the Box-Jenkins approach. George Box and Gwilym Jenkins (1976) proposed a three-stage approach to selecting the appropriate number of autoregressive and moving average components to include in a model:

1. Identification

2. Estimation

3. Diagnostic checking

To begin, it must be determined whether the data being modelled are stationary. A simple plot of the data can reveal nonstationarity produced by elements such as trending, periodicity, and structural breaks. If the data are not stationary, they are transformed so that they are—most commonly, the data are first differenced or seasonally differenced (discussed in Chapter 3 and to be discussed further in Chapter 6) or detrended/deseasonalized (as discussed in Chapter 3).

Box-Jenkins Approach: Identification

Once the data are determined to be stationary or transformed accordingly, the first step of the Box-Jenkins approach is to identify the appropriate autoregressive and moving average components to include in the model. The Box-Jenkins approach to doing this makes use of the autocorrelation and partial correlation functions of the observed time series.

In Chapter 2, we discussed the autocorrelations of a time series process and gave an example of an estimated autocorrelation function (ACF). Recall that under the assumption of stationarity, the autocorrelation at lag s is defined as follows:

$$\rho_s \equiv \frac{E\left(y_t - \mu_y\right)E\left(y_{t-s} - \mu_y\right)}{E\left(y_t - \mu_y\right)^2} \tag{5.2.5}$$

The autocorrelations are denoted as ρ_s, where s indicates the "order" of the autocorrelation; for example, $\rho_1 = \text{Corr}(y_t, y_{t-1})$, $\rho_2 = \text{Corr}(y_t, y_{t-2})$, ... , $\rho_s = \text{Corr}(y_t, y_{t-s})$. Also, recall that the ACF is a plot of the autocorrelations for a range of lags, beginning with a lag of 1. Plotting ρ_s against s is called the ACF or correlogram. Calculating the autocorrelations requires knowledge of the series mean and variance. Recall from Chapter 2 that these are not known for the true data-generating process but can be estimated from the sample data, assuming that the series is stationary or has been transformed accordingly:

$$\bar{\rho}_s = \frac{\sum_{t=s+1}^{T}(y_t - \bar{y})(y_{t-s} - \bar{y})}{\sum_{t=1}^{T}(y_t - \bar{y})^2} \qquad (5.2.6)$$

These estimated autocorrelations are useful in identifying the order of the autoregressive and moving average components of an ARMA process. The pattern of the estimated autocorrelations in the ACF can tell us something about the time series process. For example, the AR(1) data-generating process is:

$$y_t = \alpha_0 + \alpha_1 y_{t-1} + \varepsilon_t. \qquad (5.2.7)$$

For this time series process, $\rho_1 = \alpha_1$ and ρ_2, the correlation between y_t and y_{t-2}, is equal to the correlation between y_t and y_{t-1} multiplied by the correlation between y_{t-1} and y_{t-2}. From Equation 5.2.7,

$$y_{t-1} = \alpha_0 + \alpha_1 y_{t-2} + \varepsilon_{t-1}.$$

Therefore, $\rho_2 = \alpha_1 \times \alpha_1 = \alpha_1^2$. Generally, the AR(1) time series process has the following autocorrelations for $s \geq 1$:

$$\rho_s = \alpha_1^s. \qquad (5.2.8)$$

The process is assumed to be stationary, $|\alpha_1| < 1$, and so the ACF will converge to 0 exponentially, either directly if α_1 is positive or through a dampened oscillatory path if α_1 is negative. For example, consider the ACFs for data sampled from the two autoregressive data-generating processes shown in Figure 5.1. For each data-generating process, the figure includes both the theoretical ACF and the ACF for a single realization.

If we were to estimate the ACF for a time series with an unknown data-generating process and it demonstrated one of these versions of exponential convergence to 0, we might identify the time series process as AR(1).

The statistical significance of the autocorrelations can be tested based on whether they exceed 2 standard deviations, where the standard deviation is estimated from the sample series:

$$\text{Var}(\rho_s) = T^{-1}, \text{ for } s = 1.$$

$$\text{Var}(\rho_s) = T^{-1}\left(1 + 2\sum_{j=1}^{s-1}\rho_j^2\right), \text{ for } s > 1. \qquad (5.2.9)$$

Figure 5.1   Autocorrelation Functions for Autoregressive Processes

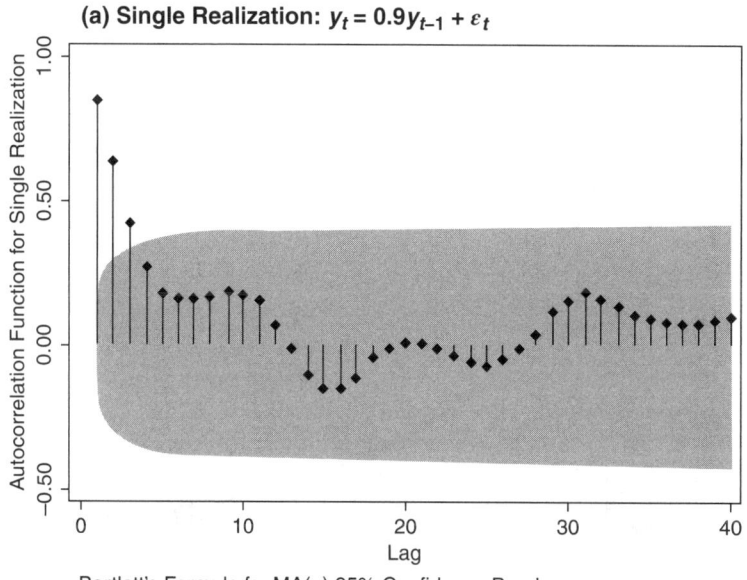

(a) Single Realization: $y_t = 0.9y_{t-1} + \varepsilon_t$

Bartlett's Formula for MA(q) 95% Confidence Bands

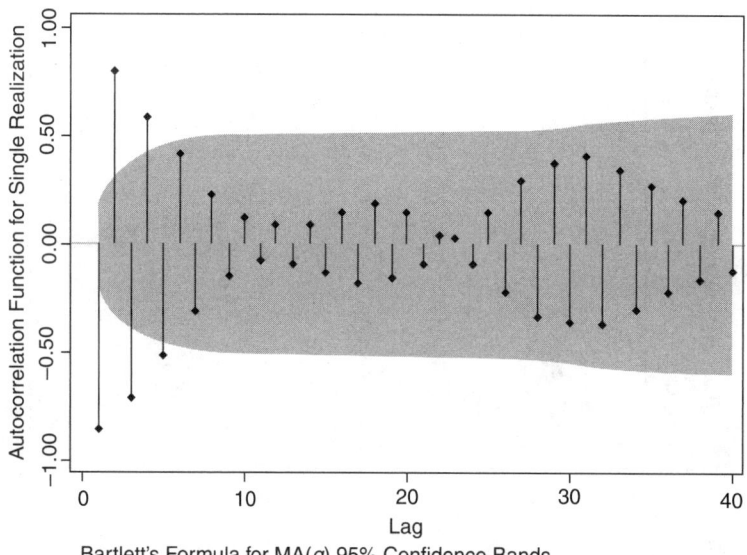

(b) Single Realization: $y_t = -0.9y_{t-1} + \varepsilon_t$

Bartlett's Formula for MA(q) 95% Confidence Bands

(Continued)

120

Figure 5.1 (Continued)

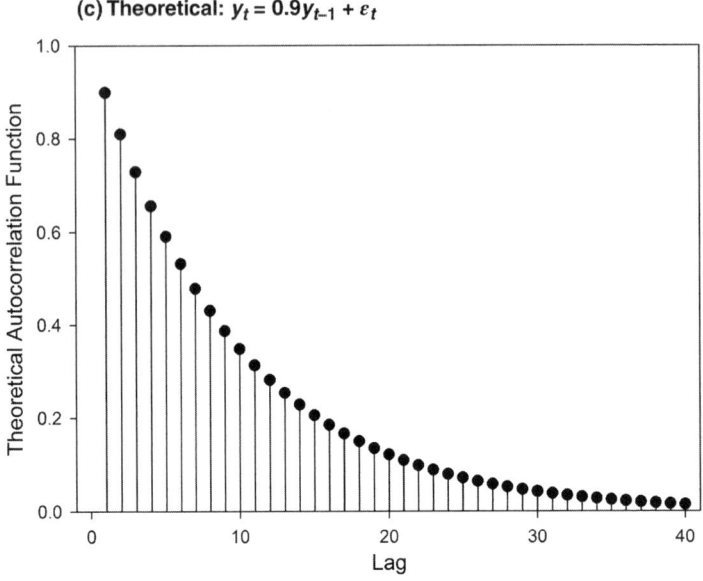

(c) Theoretical: $y_t = 0.9y_{t-1} + \varepsilon_t$

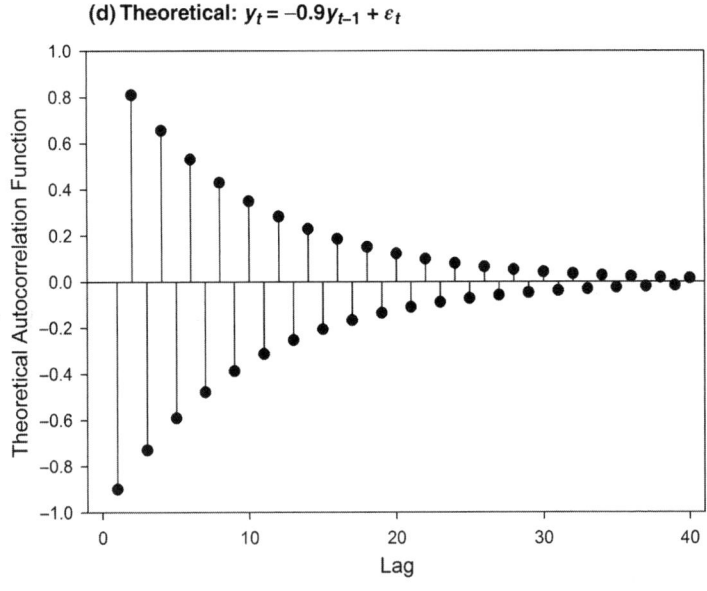

(d) Theoretical: $y_t = -0.9y_{t-1} + \varepsilon_t$

NOTE: AR = autoregressive, MA = moving average.

Again, we do not know ρ_s, but we can use their estimates ($\hat{\rho}_s$) in the above equation. The confidence envelope for the ACF in Figure 5.1 is calculated using these estimates of the variance. The null hypothesis of the test is that the autocorrelation is not significantly different from 0. This is not really intended to test any individual autocorrelation. The intent is to test whether the autocorrelations are, as a whole, significantly different from what we would expect from a white noise process. In a white noise process, only 1 in 20 will be significant at the 0.05 significance level.

For an AR(1) process other than ρ_1, the autocorrelations are *indirect* correlations. By this, we mean that there is a correlation between y_t and y_{t-2} only because y_t and y_{t-1} are correlated and y_{t-1} and y_{t-2} are correlated. It is also possible to calculate the correlation between, for example, y_t and y_{t-2} controlling for, or partialing out, the effects of the intervening values of y_{t-1}. This can be done by regressing y_t on y_{t-1} and y_{t-2}, and using the coefficient on y_{t-2} as our estimate of the partial autocorrelation φ_2. Plotting these partial correlations, φ_s, against s is the partial autocorrelation function (PACF). We can estimate the PACF and employ this in the identification of an unknown time series process.

For an AR(1) process $\varphi_1 = \rho_1$ and for $s > 1$, $\varphi_s = 0$. Consider the PACFs for the same data for which we calculated the ACFs, as shown in Figure 5.2. The variance used to calculate the confidence envelopes for these partial autocorrelations is $\text{Var}(\varphi_s) = T^{-1}$.

A PACF with this pattern would help us confirm that our data were generated by an AR(1) process. Generally, the patterns exhibited in the ACF and PACF of our time series data can be used to identify the process that generated them (Box & Pierce, 1970). An ACF and PACF, like the ones we have just observed, would suggest that the data-generating process contains a single autoregressive lag of order 1.

Things get a little more complicated for higher-order autoregressive processes. For example, consider an AR($p = 1, 2$) process:

$$y_t = \alpha_0 + \alpha_1 y_{t-1} + \alpha_2 y_{t-2} + \varepsilon_t. \qquad (5.2.10)$$

The autocorrelations and partial autocorrelations will be a function of the parameters α_1 and α_2.

$$\rho_1 = \frac{\alpha_1}{(1-\alpha_2)} \qquad \varphi_1 = \alpha_1.$$
$$\rho_2 = \alpha_1 \times \rho_1 + \alpha_2 \qquad \varphi_2 = \alpha_2 \qquad (5.2.11)$$

For $s > 2$, $\rho_s = \alpha_1 \rho_{s-1} + \alpha_2 \rho_{s-2}$ and $\varphi_s = 0$.

Figure 5.2 Partial Autocorrelation Functions for Autoregressive Processes

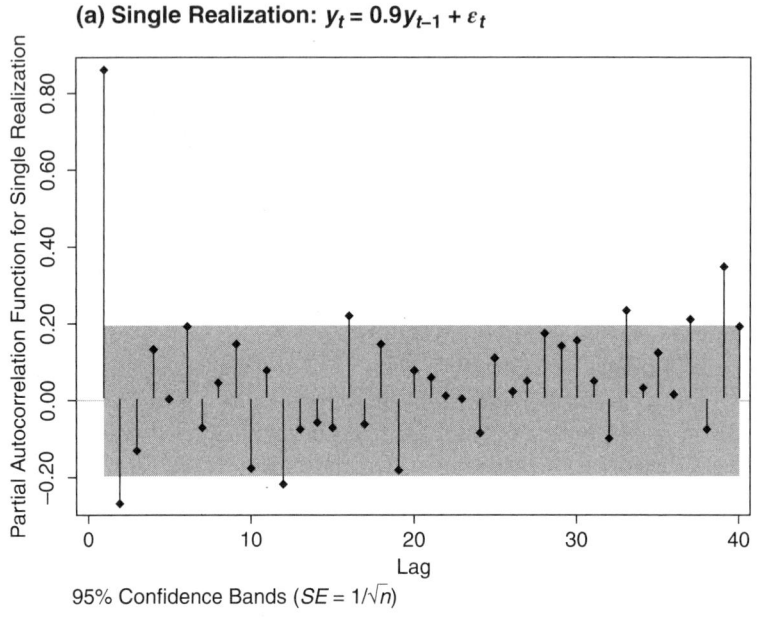

(a) Single Realization: $y_t = 0.9y_{t-1} + \varepsilon_t$

95% Confidence Bands ($SE = 1/\sqrt{n}$)

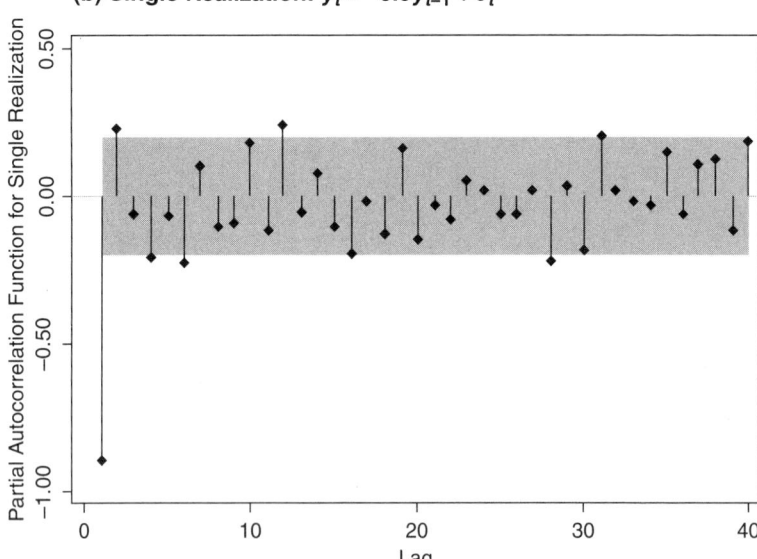

(b) Single Realization: $y_t = -0.9y_{t-1} + \varepsilon_t$

95% Confidence Bands ($SE = 1/\sqrt{n}$)

Figure 5.2

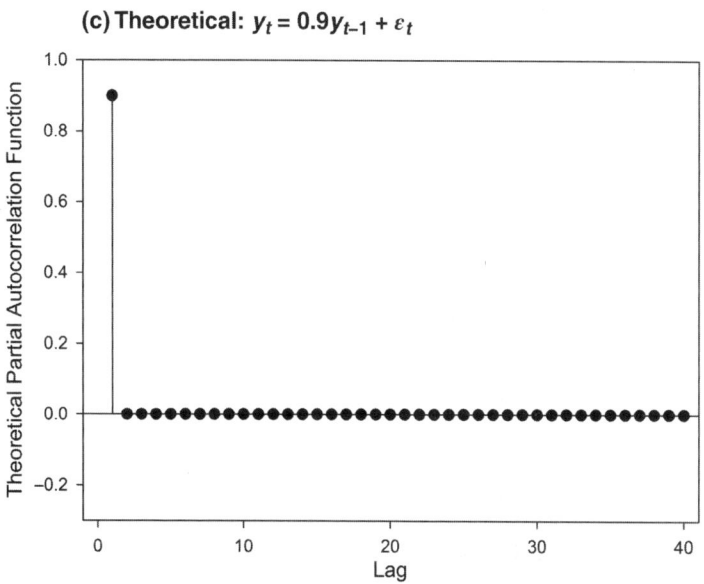

(c) Theoretical: $y_t = 0.9y_{t-1} + \varepsilon_t$

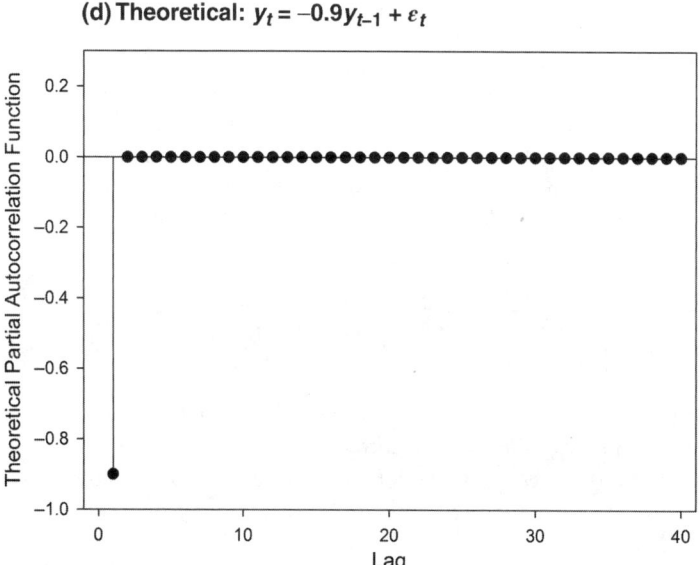

(d) Theoretical: $y_t = -0.9y_{t-1} + \varepsilon_t$

NOTE: *SE* = standard error.

Consider an AR($p = 1$, 2) data-generating process with $\alpha_1 = 0.5$ and $\alpha_2 = 0.3$:

$$y_t = 0.5y_{t-1} + 0.3y_{t-2} + \varepsilon_t.$$

$$\rho_1 = \frac{0.5}{(1 - 0.3)} = 0.7 \qquad \varphi_1 = 0.5$$

$$\rho_2 = 0.5 \times 0.7 + 0.3 = 0.65 \qquad \varphi_2 = 0.3.$$

For $s > 2$, $\varphi_s = 0$, and the ACF decays at a rate slower than that of an AR(1) process.

We have seen what the ACF and PACF look like for an autoregressive process, but the time series process may also contain moving average components. Consider the MA(1) data-generating process:

$$y_t = \varepsilon_t + \phi_1 \varepsilon_{t-1}. \tag{5.2.12}$$

It can be shown that for the ACF,

$$\rho_1 = \frac{\phi_1}{(1 + \phi_1^2)} \text{ and } \rho_s = 0 \,\forall s > 1. \tag{5.2.13}$$

It can also be shown that the PACF will decay geometrically if $\phi_1 < 0$ and will decay through an oscillatory path if $\phi_1 > 0$. Consider the ACFs and PACFs for data generated by two MA(1) processes, one with $\phi_1 = 0.9$ and the other with $\phi_1 = -0.9$, as depicted in Figures 5.3 and 5.4. With $\phi_1 = 0.9$ in the data-generating process, the first autocorrelation is

$$\rho_1 = \frac{\phi_1}{(1 + \phi_1^2)} = \frac{0.9}{(1 + 0.9^2)} = 0.5,$$

which is what we observe for $\hat{\rho}_1$ in the ACF. With $\phi_1 = -0.9$ in the data-generating process, the first autocorrelation is -0.5, which again is what we observe for $\hat{\rho}_1$ in the ACF. As for the PACFs, they decay approximately geometrically—directly with $\phi_1 = -0.9$ and through an oscillating path when $\phi_1 = 0.9$.

Generally, the pattern of the ACF and PACF indicates the order of p and q in the data-generating process. This can be used to inform the order of p and q required in our ARMA(p,q) model in order to capture the time series dynamics.

Figure 5.3   Autocorrelation Functions for Moving Average Processes

(a) Single Realization: $y_t = \varepsilon_t + 0.9\varepsilon_{t-1}$

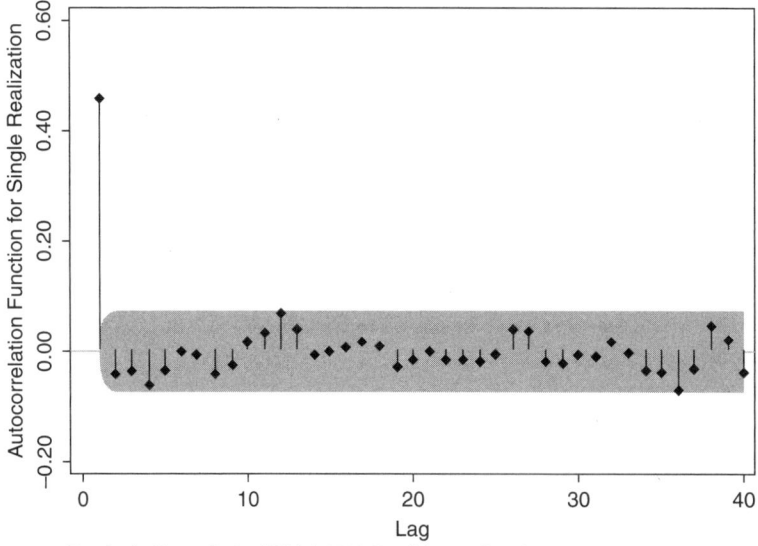

Bartlett's Formula for MA(q) 95% Confidence Bands

(b) Single Realization: $y_t = \varepsilon_t - 0.9\varepsilon_{t-1}$

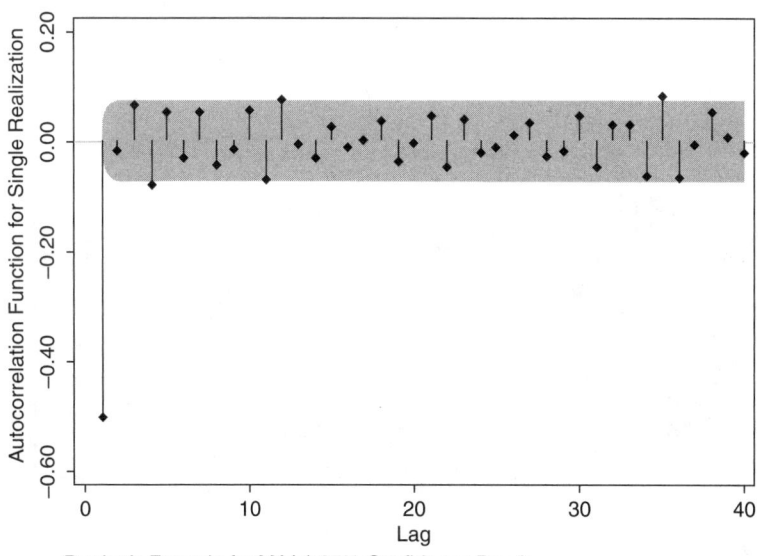

Bartlett's Formula for MA(q) 95% Confidence Bands

(Continued)

126

Figure 5.3 (Continued)

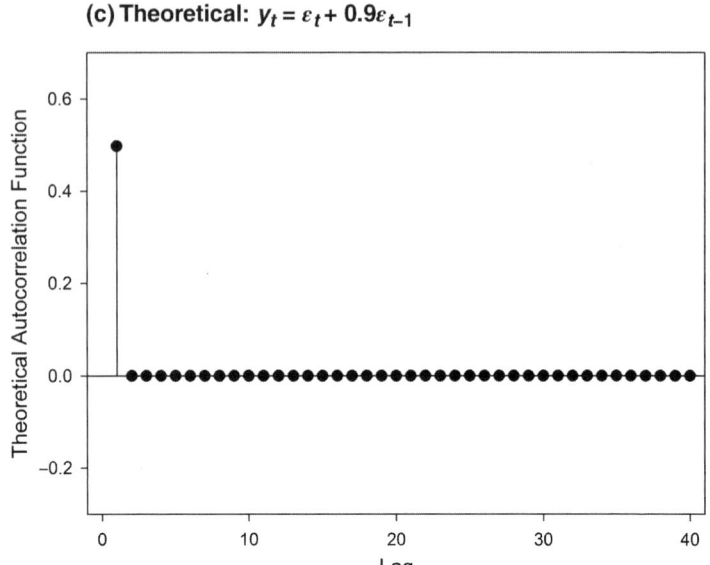

(c) Theoretical: $y_t = \varepsilon_t + 0.9\varepsilon_{t-1}$

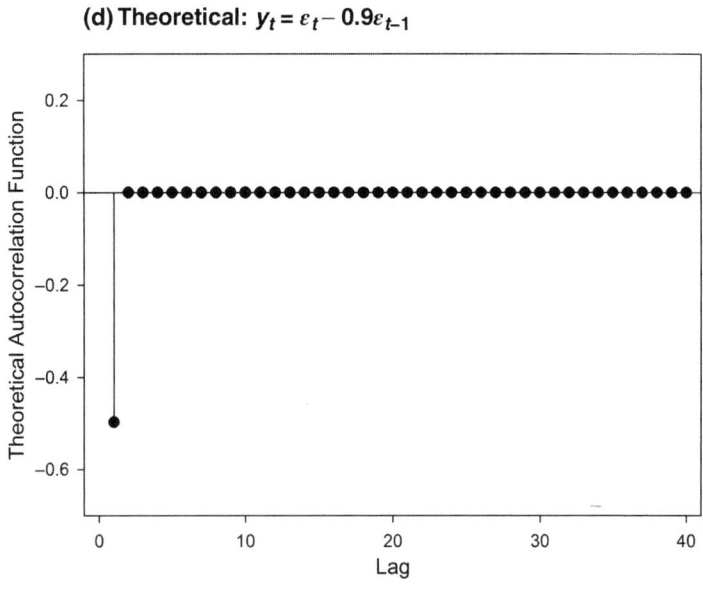

(d) Theoretical: $y_t = \varepsilon_t - 0.9\varepsilon_{t-1}$

NOTE: MA = moving average.

Figure 5.4 Partial Autocorrelation Functions for Moving Average Processes

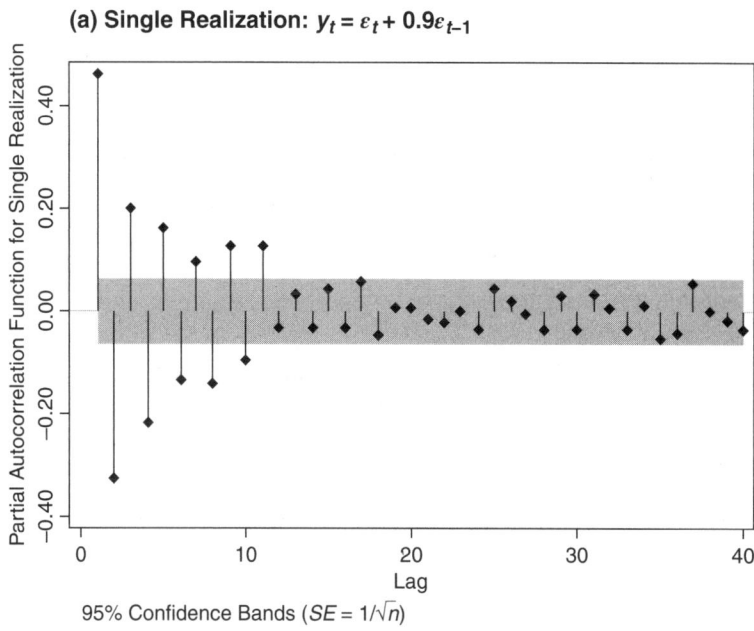

(a) Single Realization: $y_t = \varepsilon_t + 0.9\varepsilon_{t-1}$

95% Confidence Bands ($SE = 1/\sqrt{n}$)

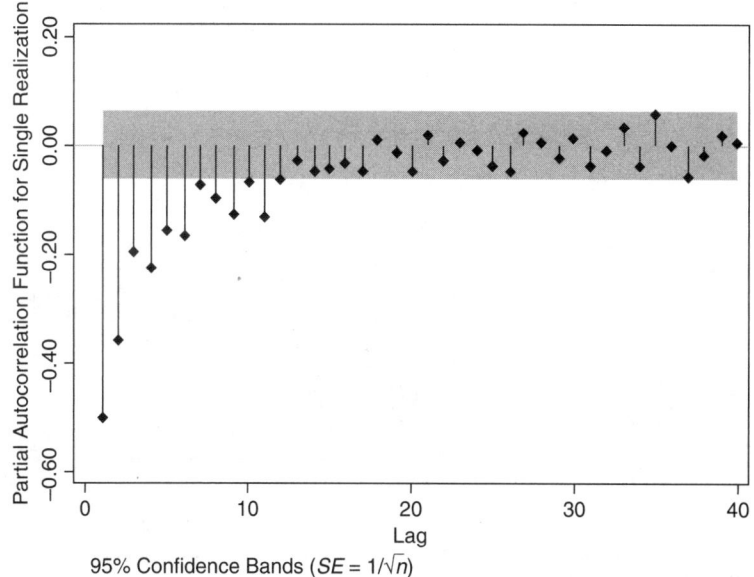

(b) Single Realization: $y_t = \varepsilon_t - 0.9\varepsilon_{t-1}$

95% Confidence Bands ($SE = 1/\sqrt{n}$)

(Continued)

128

Figure 5.4   (Continued)

(c) Theoretical: $y_t = \varepsilon_t + 0.9\varepsilon_{t-1}$

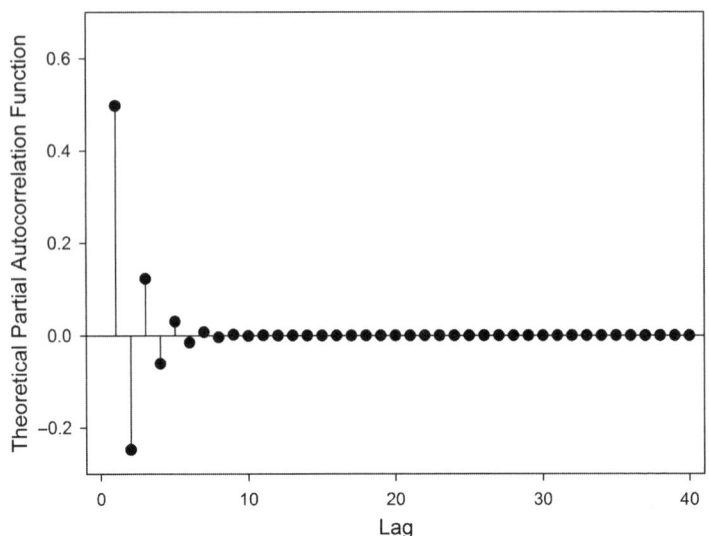

(d) Theoretical: $y_t = \varepsilon_t - 0.9\varepsilon_{t-1}$

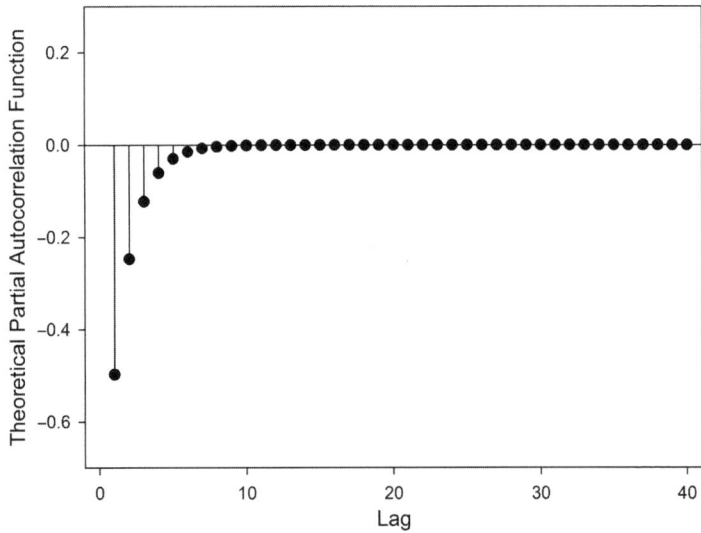

NOTE: *SE* = Standard Error.

Some basic guidelines for identifying some common processes based on their ACF and PACF are described in Table 5.1 (Enders, 2004, pg. 66).

For complex processes, the correct combination of p and q may not be clear. In fact, more than one combination may fit the same data-generating process. In selecting the best ARMA data model, Box-Jenkins set out model selection criteria. The goal is to select a data model that accounts for the autocorrelation in the error term (i.e., once we include autoregressive and moving average terms, we are left with residuals that are a white noise process) while following the principle of parsimony. We choose the model that fits the data best, but (a) if two alternative models fit the data approximately equally well, we choose the model with fewer coefficients, and (b) each coefficient should be significantly different from 0 at our chosen significance level—typically 0.05.

When testing how well a model fits the data, a number of goodness-of-fit measures are at our disposal. Some of the most common are the Akaike Information Criterion (AIC) and the Schwartz Bayesian Information Criterion (BIC). These are calculated as follows (Enders, 2004):

$$AIC = T \ln(\text{sum of squared residuals}) + 2(k),$$

$$BIC = T \ln(\text{sum of squared residuals}) = ln(T)(k). \qquad (5.2.14)$$

For both formulae, k is the number of parameters in the model including the intercept. Models with smaller (including more negative) AIC and BIC

Table 5.1 Guidelines for Identifying Autoregressive and Moving Average Processes

Process	ACF	PACF
White noise	All $\rho_s = 0$, $s \neq 0$.	All $\varphi_s = 0$.
AR($p = P$)	Decay toward zero. Coefficients may oscillate (decay is geometric if $P = 1$).	Spikes at $s = P$. All $\varphi_s = 0$ for $s > P$.
MA($q = Q$)	Spike at $s = Q$ and $\rho_s = 0$ $\forall s \neq Q$.	Decay toward zero (either direct or oscillatory).
ARMA ($p = P$, $q = Q$)	Decay (either direct or oscillatory) beginning at lag Q.	Decay (either direct or oscillatory) beginning after lag P.

NOTE: ACF = autocorrelation function, AR = autoregressive, ARMA = autoregressive moving average, MA = moving average, PACF = partial autocorrelation function.

values fit better. The BIC, which applies a greater penalty for additional parameters, will select a more parsimonious model and has better properties when T is large. The AIC, however, may be superior with a small T (Harvey, 1993).

When comparing BIC values from two different models, we can use the rough guidelines given in Table 5.2 (Raftery, 1995) to decide if there is evidence that one fits better than the other.

We can also use a likelihood ratio (LR) test to compare two nested models.

$$LR = -2\left(\text{Likelihood for null model} - \text{Likelihood for alternative model}\right).$$

This statistic has a chi-squared distribution, with the degrees of freedom equal to the difference in the number of parameters between the two nested models, and so the usual methods of inference can be used.

It is important to keep T constant when comparing models. This should be kept in mind as including an additional autoregressive lag, in practice, means eliminating a data point. This data point should also be eliminated for all the models that are being considered.

When comparing models, it is important to always keep in mind that the calculation of the ACF and PACF is based on the assumption that the sequence $\{y_t\}$ is stationary. We should therefore be suspicious of any model that produces estimates of the coefficients that suggest instability (e.g., for an AR(1) model, $|\alpha_1| \geq 1$).

Let us now consider an example of model identification using the Box-Jenkins approach for time series data with an unknown data-generating process. For this example, we shall look at monthly vote intention data for the Canadian federal government (the proportion of survey respondents who indicated that they would vote for the party currently in government if an election were held tomorrow). The period we will examine is 1993 to 2000,

Table 5.2 Guidelines for Comparing Models Using BIC Values

Difference	Evidence
0–2	Weak
2–6	Positive
6–10	Strong
>10	Very strong

NOTE: BIC = Schwartz Bayesian Information Criterion.

during which there was a Liberal federal government (Figure 5.5). We will also want to include exogenous regressors in our model. We are interested in determining whether economic conditions affected vote intention for the governing party (gross domestic product [GDP], inflation, and unemployment)—this will be a classic economic popularity model.

We begin by plotting the vote intention variable: the proportion of respondents indicating that they would vote for the Liberal governing party if an election were held.

Visually, there is some evidence of trending, and (more important) there are theoretical reasons to believe that there is a downward trend. After an initial honeymoon, new Canadian governments generally lose popularity over their term in office as they are forced to make difficult choices that inevitably upset some of those who voted for the governing party. As demonstrated in Chapter 2, we can test for the trend by regressing the vote intention data on a time variable. This produces the results presented in Table 5.3.

We can see from the results that although the magnitude of the trend is small, it is statistically significant at the 0.05 significance level. Next, we may choose to produce a detrended vote intention variable by using the

Figure 5.5   Vote Intention for the Liberal Party of Canada, 1993 to 2000

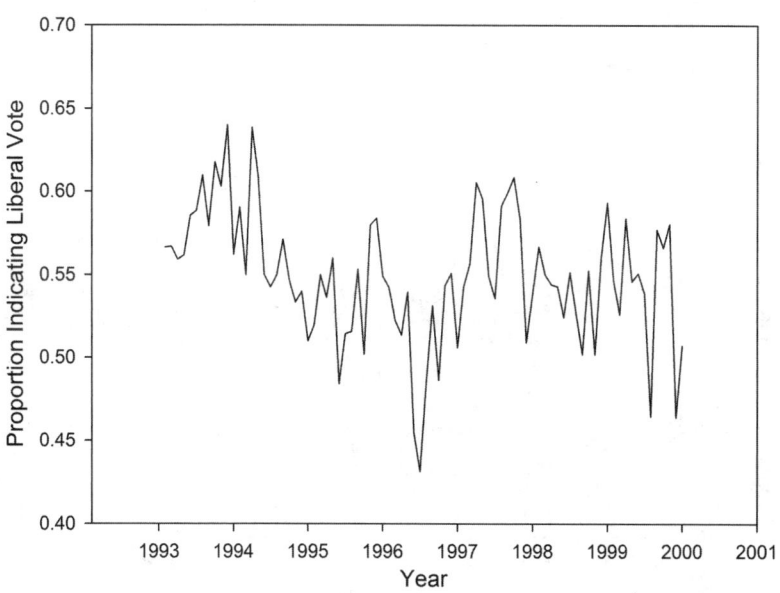

estimated equation to calculate the residuals from the above regression. The resulting time series is the vote intention time series with the trend partialed out, as we can see in the plot of this time series shown in Figure 5.6.

As we will discuss in Chapter 6, we might ask ourselves whether the original data are integrated (defined in Chapter 2), rather than containing a deterministic linear trend. We will leave this issue for now and assume that they are not and that they are stationary once the deterministic linear trend is removed.

Table 5.3   Trending in Vote Intention for the Incumbent Government in Canada

Vote	Coefficient	Standard Error	t Statistic	P Value
Trend	−0.00046	0.00017	−2.63	0.010
Intercept	0.57	0.0085	67.902	<0.001

NOTE: $R^2 = 0.078$, $T = 84$; T = number of time points.

Figure 5.6   Detrended Vote Intention for the Liberal Party of Canada, 1993 to 2000

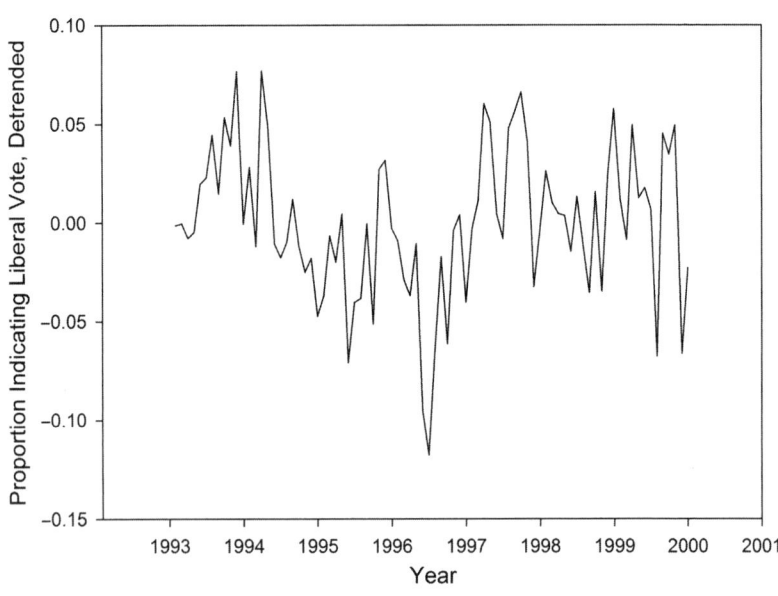

Having addressed the issue of stationarity, we next estimate and examine the ACF and PACF for the detrended vote intention variable. These are displayed in Figure 5.7.

Examining the ACF, it appears that the autocorrelations begin to decay at $s = 1$ and again at $s = 4$. Meanwhile, the PACF has spikes at $s = 1$ and $s = 4$. The PACF also appears to have spikes at $s = 7$ and higher, but there are no corresponding significant autocorrelations in the ACF. As a general rule, it is best to focus on the autocorrelations and partial autocorrelations at the lowest s, and worry about the higher values of s only if we are unable to come up with an ARMA model that accounts for all autocorrelation within the residuals. This is based on the principle of parsimony, which is key to the Box-Jenkins approach.

The patterns found in the ACF and PACF suggest that we might try including an AR(1) and an AR($p = 4$) term in our model. We would denote this as an ARMA ($p = 1,4$) model. Having identified the ARMA($p = 1,4$) model that we think will capture the time series process, the next step is estimation.

Figure 5.7   Autocorrelation and Partial Autocorrelation Functions for Detrended Vote Intention

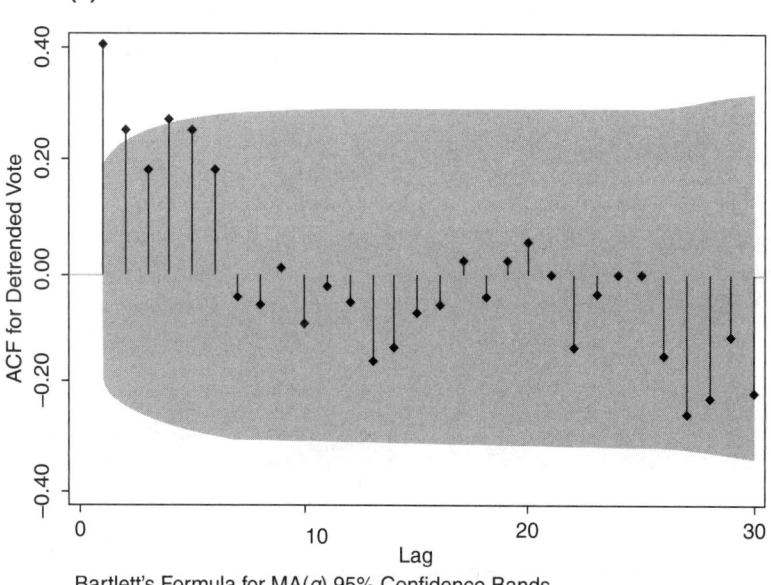

(a) Autocorrelation Function

Bartlett's Formula for MA(q) 95% Confidence Bands

(Continued)

Figure 5.7 (Continued)

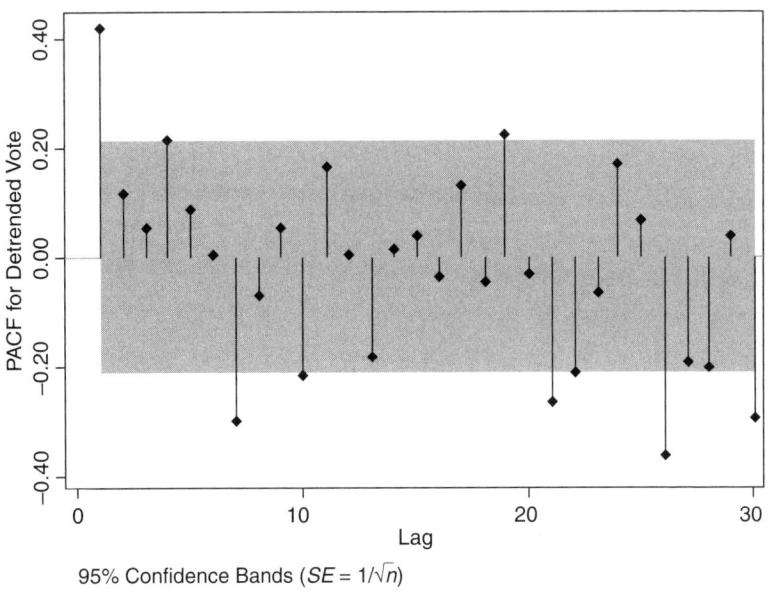

(b) Partial Autocorrelation Function

NOTE: ACF = autocorrelation function, MA = moving average, PACF = partial autocorrelation function, *SE* = standard error.

Box-Jenkins Approach: Estimation

Once an ARMA model is selected, the α_is and ϕ_js can be estimated. This is the second stage of the Box-Jenkins approach. Estimation is done using maximum likelihood estimation. Recall the two-component representation of the ARMA process:

$$y_t = \beta_0 + \mu_t,$$

$$\mu_t = \sum_{i=1}^{p} \alpha_i \mu_{t-i} + \sum_{j=1}^{q} \phi_j \varepsilon_{t-j} + \varepsilon_t, \qquad (5.2.15)$$

where the first equation is called the structural component and the second is the disturbance component. This is the representation often used for the purposes of model estimation and interpretation of the estimation results. The two-component representation of the ARMA($p = 1,4$) data-generating process is

$$y_t = \beta_0 + \mu_t,$$

$$\mu_t = \alpha_1 \mu_{t-1} + \alpha_4 \mu_{t-4} + \varepsilon_t. \tag{5.2.16}$$

Applied to the vote intention data, the maximum likelihood estimates of the data model parameters are presented in Table 5.4.

We can see from the estimation results that both AR(1) and AR(4) coefficients are statistically significant at the 0.05 significance level. The next step is diagnostics.

Box-Jenkins Approach: Diagnostic Checking

The first diagnostic check is to determine whether the estimated errors are free from serial correlation or any other unwanted patterns. In other words, we test the residuals to determine if they are a white noise process. This can be done using the Portmanteau Q statistic for white noise, along with the ACF and PACF. The Q statistic is 47.91 and is chi-squared distributed with 40 degrees of freedom. This gives us a P value of 0.18. We are not able to reject the null hypothesis of a white noise process for the residuals. This suggests that the estimated errors do not include serial correlation, trending, or periodicity. The ACF and PACF of the estimated errors can also help us check if the correct model was specified (Figure 5.8).

We can see from the PACF that there may still be some autocorrelation in the seventh lag. Based on this information, we may try estimating an ARMA($p = 1,4,7$) model (Table 5.5).

Table 5.4 Liberal Government Vote Intention—AR(1,4)

Vote	Coefficient	Standard Error	z Statistic	P Value
AR(1)	0.37	0.090	4.15	<0.001
AR(4)	0.21	0.088	2.42	0.015
Intercept	−0.00036	0.0091	−0.04	0.969

NOTE: Log likelihood = 165.7978, $T = 84$; T = number of time points, AR = autoregressive.

Figure 5.8 Autocorrelation and Partial Autocorrelation Functions of Liberal Government Vote Model Residuals

(a) Autocorrelation Function

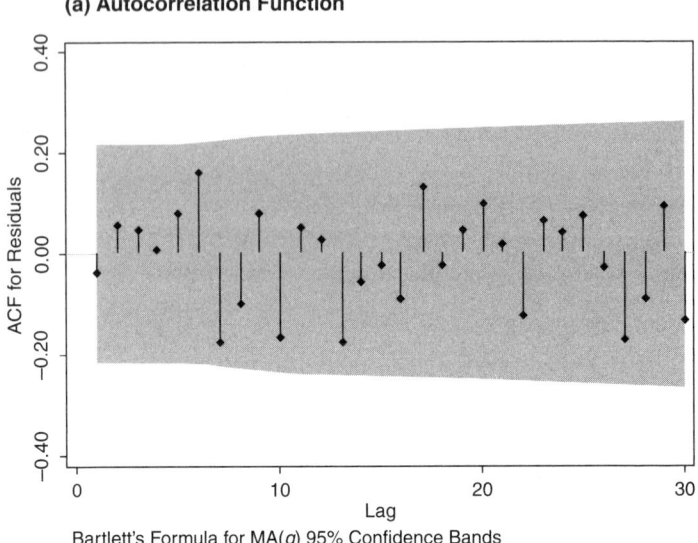

Bartlett's Formula for MA(*q*) 95% Confidence Bands

(b) Partial Autocorrelation Function

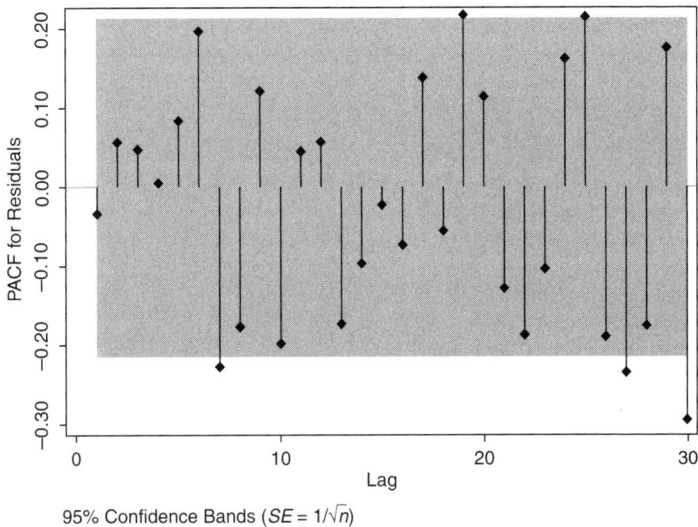

95% Confidence Bands ($SE = 1/\sqrt{n}$)

NOTE: ACF = autocorrelation function, MA = moving average, PACF = partial autocorrelation function, SE = standard error.

Table 5.5 Liberal Government Vote Intention—AR(1,4,7)

Vote	Coefficient	Standard Error	z Statistic	P Value
AR(1)	0.40	0.090	4.48	<0.001
AR(4)	0.26	0.095	2.75	0.006
AR(7)	−0.20	0.11	−1.75	0.081
Intercept	−0.00056	0.0068	−0.08	0.935

NOTE: Log likelihood = 167.678, $T = 84$; T = number of time points, AR = autoregressive.

Table 5.6 Liberal Government Vote Intention—AR(1,4) MA(7)

Vote	Coefficient	Standard Error	z Statistic	P Value
AR(1)	0.39	0.091	4.35	<0.001
AR(4)	0.25	0.096	2.61	0.009
MA(7)	−0.25	0.12	−2.13	0.033
Intercept	−0.00028	0.0077	−0.040	0.971

NOTE: Log likelihood = 167.742, $T = 84$; T = number of time points, AR = autoregressive, MA = moving average.

Looking at these results, we see that the AR(7) term is not statistically significant at the 0.05 significance level. Based on the Box-Jenkins approach, we would not include this term. Looking at the previous PACF (Figure 5.7), it is not clear whether the autocorrelation at the seventh lag is due to an AR(7) or an MA(7) term. On this basis, we might try estimating an ARMA($p = 1, 4, q = 7$) model (Table 5.6).

All terms in this model are significant at the 0.05 significance level. Again, we can test the residuals against the null of a white noise process and examine their ACF and PACF. The Q statistic is 43.64, and the corresponding P value is 0.32, again indicating no evidence that serial correlation remains within the residuals.

The ACF and PACF for the residuals (Figure 5.9) show no signs of any remaining autocorrelation. It appears that we have specified a model that correctly accounts for the autocorrelation in the residuals. If autocorrelations

Figure 5.9 Autocorrelation and Partial Autocorrelation Functions of Liberal Government Vote Model Residuals

(a) Autocorrelation Function

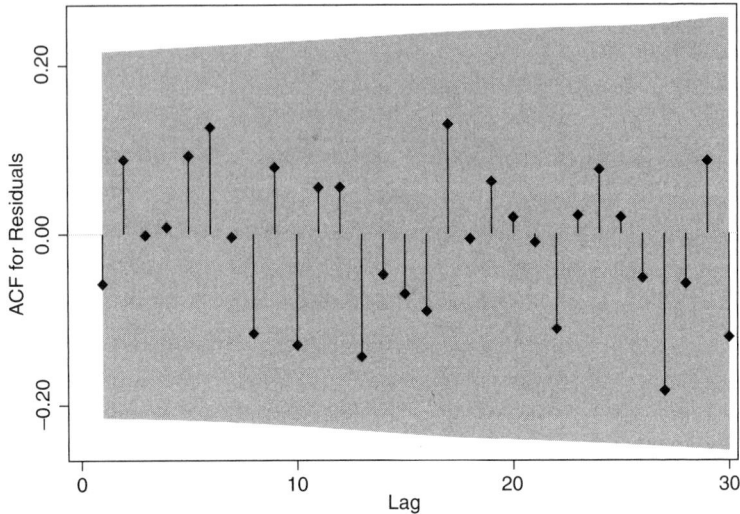

Bartlett's Formula for MA(q) 95% Confidence Bands

(b) Partial Autocorrelation Function

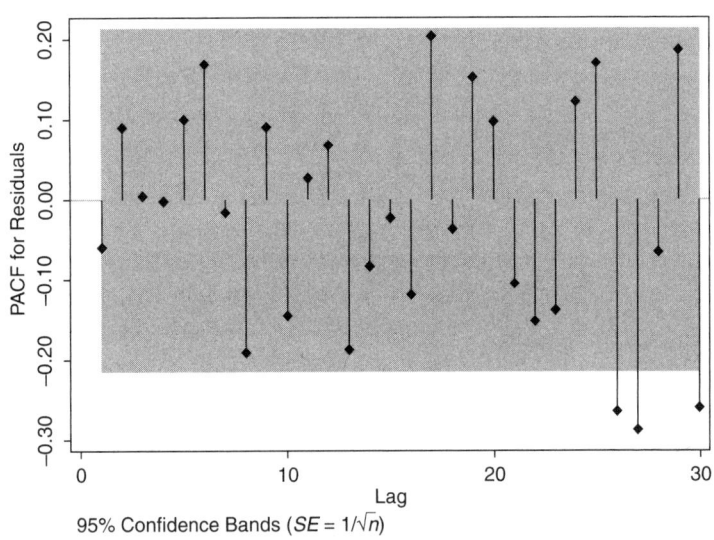

95% Confidence Bands ($SE = 1/\sqrt{n}$)

NOTE: ACF = autocorrelation function, MA = moving average, PACF = partial autocorrelation function, SE = standard error.

are still found in the estimated errors, the model can again be respecified and reestimated.[1]

Finally, we may want to compare the fit of ARMA($p = 1, 4, q = 7$) with the simpler ARMA($p = 1, 4$). These models are nested, and so we can do this with the LR test. We get an LR value of 3.89. This is chi-squared distributed with 1 degree of freedom, giving us a P value of 0.049. We can reject the null hypothesis that the ARMA($p = 1, 4, q = 7$) model fits no better than the ARMA($p = 1, 4$) model.

5.3 Autoregressive Moving Average (ARMA) Models With Exogenous Regressors

Independent variables (exogenous regressors) and/or their lags can be added to the first component of the two-component representation of the ARMA model:

$$y_t = \beta_0 + \sum_{j=1}^{k} \beta_j x_j + \mu_t,$$

$$\mu_t = \sum_{i=1}^{p} \alpha_i \mu_{t-i} + \sum_{j=1}^{q} \phi_j \varepsilon_{t-j} + \varepsilon_t, \qquad (5.3.1)$$

where $\sum_{j=1}^{k} \beta_j x_j$ represents the exogenous regressors (including lags) that we wish to include in the model. Returning to our example, we want to include economic conditions as variables. Specifically, we include the third lag (recall that the data are monthly) of year-over-year change in GDP and inflation and the sixth lag of unemployment. The lags included are based on theoretical considerations (e.g., the timing of the release of such economic figures). This produces the following estimates (Table 5.7).

We see from the results that the AR(1) and AR(4) terms are no longer statistically significant. This sometimes happens once the independent variables are included. This may also occur because we have lost six data points by including the sixth lag of unemployment. Given the Box-Jenkins guidelines to include only the autoregressive and moving average terms that reach statistical significance, we may now want to rerun the model without

---

[1] The tests of the null hypothesis of no skewness and of the null hypothesis of no kurtosis (relative to the normal) have P values of 0.2807 and 0.699, respectively. We cannot reject the null hypotheses that the skewness and kurtosis of the residuals do not deviate from what is expected for a normal distribution.

the AR(1) and AR(4) terms and then test the errors. We might also want to test if either of these terms is statistically significant when entered individually. If we did this, we would find that they are not. The results from the model without these terms are given in Table 5.8.

We can now predict the errors and calculate the Q statistic to test if they follow a white noise process. The Q statistic is 41.26, and the corresponding

Table 5.7   Canadian Economic Voting Model—AR(1,4) MA(7)

Vote	Coefficient	Standard Error	z Statistic	P Value
L3. GDP	0.0085	0.0034	2.51	0.012
L3. Inf	−0.023	0.0065	−3.53	<0.001
L6. Unemployment	−0.0053	0.0051	−1.05	0.296
Constant	0.059	0.059	1.00	0.319
AR(1)	0.19	0.11	1.63	0.103
AR(4)	0.083	0.15	0.56	0.573
MA(7)	−0.33	0.12	−2.71	0.007

NOTE: Log likelihood = 161.704, T = 78; T = number of time points, AR = autoregressive, GDP = gross domestic product, MA = moving average, L3 = third lag, L6 = sixth lag.

Table 5.8   Canadian Economic Voting Model—MA(7)

Vote	Coefficient	Standard Error	z Statistic	P Value
L3. GDP	0.0082	0.0027	3.05	0.002
L3. Inf	−0.023	0.0052	−4.37	<0.001
L6. Unemployment	−0.0057	0.0039	−1.47	0.143
Constant	0.063	0.045	1.38	0.168
MA(7)	−0.38	0.12	−3.22	0.001

NOTE: Log likelihood = 160.22, T = 78; T = number of time points, GDP = gross domestic product, MA = moving average, L3 = third lag, L6 = sixth lag.

P value is 0.29. Based on these results, we cannot reject the null hypothesis that the errors follow a white noise process. We can also examine the ACF and PACF for the residuals (Figure 5.10).

No evidence of autocorrelation remaining in the errors can be found in the ACF. The PACF shows little sign of autocorrelation. This evidence, in combination with the *Q* test, indicates that we have specified a model that correctly accounts for the autocorrelation in the residuals.

Again, we may want to compare model fits. Specifically, we might want to compare the ARMA($p = 1, 4, q = 7$) model without exogenous regressors with the ARMA($q = 7$) with exogenous regressors. We might do this using BIC values.[2] For a fair comparison, we must compare the BIC values for the two models estimated on the same data. This requires us to reestimate the ARMA($p = 1, 4, q = 7$) model without the first six points. The BIC for this model is −284.29, and for the ARMA($q = 7$) model with exogenous

Figure 5.10   Autocorrelation and Partial Autocorrelation Functions for Residuals

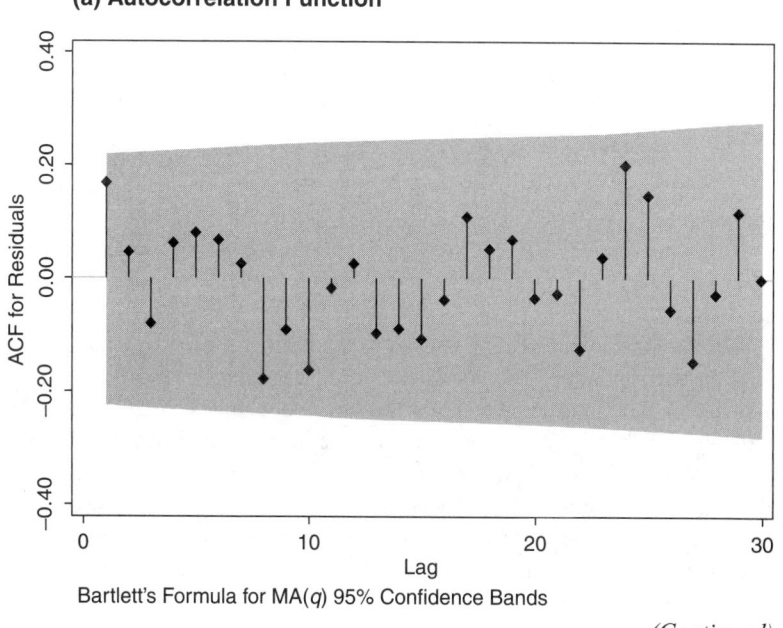

(a) Autocorrelation Function

Bartlett's Formula for MA(*q*) 95% Confidence Bands

(Continued)

---

[2] The LR test is inappropriate here as the models are not nested.

Figure 5.10 (Continued)

(b) Partial Autocorrelation Function

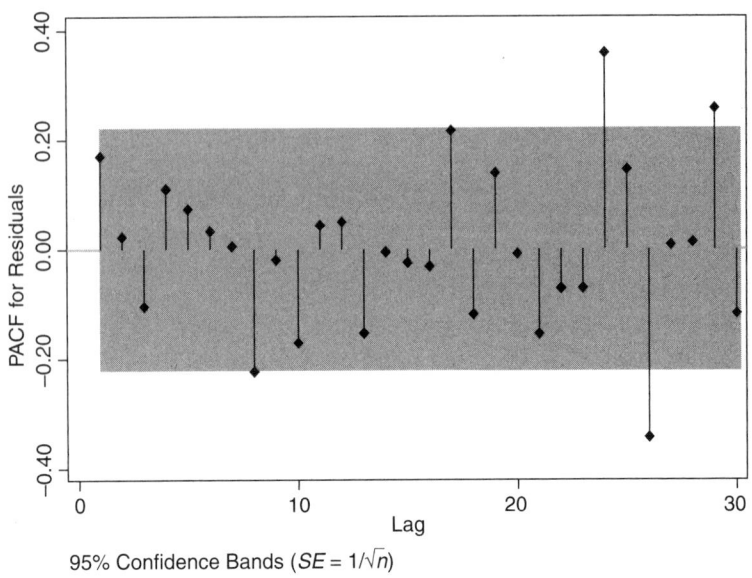

95% Confidence Bands ($SE = 1/\sqrt{n}$)

NOTE: ACF = autocorrelation function, MA = moving average, PACF = partial autocorrelation function, SE = standard error.

regressors, it is −294.31. The BIC for the model with exogenous regressors is 10 smaller than the BIC for the model without. Referring to Table 5.2, this is strong to very strong evidence that the model with exogenous regressors fits better.

We now turn to interpreting the estimated effects of the exogenous regressors on vote intention. The coefficients on the independent variables in the ARMA are the long-run effects of a permanent one-unit increase in the independent variable. However, they are also the immediate effects. Just like the static model in Chapter 3, the independent variables in an ARMA model have their full effect immediately. This is different from an LDV model where the independent variables have an initial short-run effect, which then builds into a long-run effect. We can see why this is so by considering the ARMA(1,1) process with a single exogenous regressor:

$$y_t = \beta_0 + \beta_1 x_t + \mu_t, \tag{5.3.2}$$

$$\mu_t = \alpha_1 \mu_{t-1} + \phi_1 \varepsilon_{t-1} + \varepsilon_t. \tag{5.3.3}$$

By rearranging Equation 5.3.2,

$$\mu_t = y_t - \beta_0 - \beta_1 x_t, \tag{5.3.4}$$

and lagging one period,

$$\mu_{t-1} = y_{t-1} - \beta_0 - \beta_1 x_{t-1}. \tag{5.3.5}$$

We can now insert Equations 5.3.4 and 5.3.5 into Equation 5.3.3,

$$y_t - \beta_0 - \beta_1 x_t = \alpha_1 \left(y_{t-1} - \beta_0 - \beta_1 x_{t-1} \right) + \phi_1 \varepsilon_{t-1} + \varepsilon_t,$$

and rearrange the terms:

$$y_t = \alpha_1 y_{t-1} - \alpha_1 \beta_0 - \alpha_1 \beta_1 x_{t-1} + \beta_0 + \beta_1 x_t + \phi_1 \varepsilon_{t-1} + \varepsilon_t,$$

$$y_t = \alpha_1 y_{t-1} + \left(1 - \alpha_1 \right) \beta_0 + \beta_1 x_t - \alpha_1 \beta_1 x_{t-1} + \phi_1 \varepsilon_{t-1} + \varepsilon_t. \tag{5.3.6}$$

This is the ADL(1,1) model from Chapter 4, with the addition of a moving average error and the constraint that the coefficient for the first lag of x_t is equal to the autoregressive parameter times the coefficient on x_t times -1: $-\alpha_1 \beta_1$. For an ADL(1,1), the immediate effect of a one-unit increase in x_t is β_1, and the long-run effect of x_t is

$$\frac{\beta_1 - \alpha_1 \beta_1}{\left(1 - \alpha_1 \right)} = \beta_1. \tag{5.3.7}$$

The constraint on the coefficient for the first lag of x_t guarantees that the long-run effect is equal to the initial effect: β_1. We can only include a long-run effect that differs from the immediate effect in the ARMA model by adding lags of the independent variables (like the FDL model in Chapter 3).

Returning to our estimates and using the 0.05 significance level, it would appear that year-over-year growth in GDP has a small but positive effect on vote intention for the party in government, while inflation has a reasonably large and negative effect: An increase of 1 percentage point in inflation reduces government popularity by 2.3 percentage points.

We can apply the formula for calculating the equilibrium of an ADL to Equation 5.3.6:

$$\frac{\left(1 - \alpha_1 \right) \beta_0}{\left(1 - \alpha_1 \right)} = \beta_0.$$

The β_0 in the ARMA is the long-run equilibrium, when growth, inflation, and unemployment are all zero. In reality, this is not a particularly likely scenario.

Having covered the basics of the Box-Jenkins approach to identifying, estimating, and testing ARMA models, in the next chapter, we will examine data with a trend of a different sort than we have encountered so far. We will examine the possibility that our data are a unit root process and, therefore, not stable. We will extend the ARMA approach to be able to model such data—the autoregressive integrated moving average (ARIMA) model. We will also consider how to account for periodicity in our data with such models. In the following section, we will look further at including exogenous variables, as we discuss transfer functions and intervention analysis.

5.4 Interventions and Transfer Functions

In Chapter 3, we estimated a model of (the log of) the number of drivers in the United Kingdom who were killed or seriously injured (KSI) in traffic accidents, using monthly data between January 1969 and December 1984 (Harvey & Durbin, 1986). We included variables that controlled for and estimated the magnitudes of seasonality and a structural break due to the introduction of a new seatbelt law at $t = 170$ (February 1983). The variable "seatbelt law" is coded "0" before $t = 170$ and "1" at $t = 170$ and afterward. The inclusion of this variable is a basic form of intervention analysis. The seatbelt law is a policy intervention that is expected to affect KSIs. We assume that the effect is to produce a change in the mean level of the dependent variable, "KSI," at time point $t = 170$. This change is assumed to be permanent. This is called a step function. We allow the magnitude of the shift in the mean to be estimated. The results shown in Table 3.8 indicate that the magnitude of the mean shift is a 15% reduction in KSIs per month.

This type of intervention is an example of a level change (Pourahmadi, 2001) or a deterministic-step change (Tsay, 1988). Such an intervention could also include additional dynamics such as building in magnitude over time (Box & Tiao, 1975). Other types of interventions include additive outliers, innovation outliers, transient changes, or any combination of these (Pourahmadi, 2001). An additive outlier is a change in the mean at a particular time point with an immediate reversion to the original mean at the next time point. The residual effect of the outlier may last for some time if the model is dynamic. Such an intervention can be represented in a model by including a variable that is coded "0" before the intervention, "1" at the time of the intervention, and "0" afterward. This is called a pulse function.

An innovation outlier is similar to an additive outlier except that it affects the innovations and would be modelled by including a variable representing a pulse function as multiplicative heteroskedasticity in a autoregressive conditional heteroscedasticity (ARCH) process. The included variable is the same as that for additive outliers, except that it is included in the model of the error variance. A transient change is also like an additive outlier, except that the change in the mean of the series due to an intervention dissipates slowly with time. These can take many different forms, and we will examine them shortly, when we discuss transfer functions.

We could also model a structural break in the covariance between variables. For example, the effect of petrol prices on KSI may be suspected of changing after the introduction of the new seatbelt law. This can be modelled by including an interaction between the seatbelt law step function and petrol prices.

In some instances, we are uncertain regarding the exact timing or nature of an intervention. In this case, we may want to conduct a more exploratory analysis. This can be done by examining the residuals from an out-of-sample forecast, as illustrated in the following example. Let us say we believe that an intervention of some form occurred after time point 169. We begin by estimating a model using just the data up to time point 169—prior to the potential intervention.

Previously, we did not examine the ACF or PACF for the residuals to determine if any autoregressive or moving average terms should be included in the model. If we follow the Box-Jenkins approach to model specification, we might determine that the appropriate model for ln(KSI) is ARMA(1,1). To see this, estimate the model from Chapter 3 using data up to time point 169. This model includes the monthly dummy variables and the log of petrol prices. It does not include the seatbelt law variable, as this occurred after time point 169. We next examine the autocorrelation and partial autocorrelations from the resulting residuals (Figure 5.11).

The ACF suggests the presence of an AR(1) process, while the PACF suggests the presence of an MA(1) process. The possibility of an MA(1) process is suggested by the fact that the spike at a lag of 1 is followed by a shorter spike at a lag of 2 and even shorter spikes at lags of 3 and 4—that is, decaying partial autocorrelations. The results from estimating ARMA(1,1) model are given in Table 5.9.

Next we produce predicted values for ln(KSI) using the parameters from the model estimated using only the data up to time point 169.

Figure 5.11   Autocorrelation and Partial Autocorrelation Functions for
Residuals

(a) Autocorrelation Function

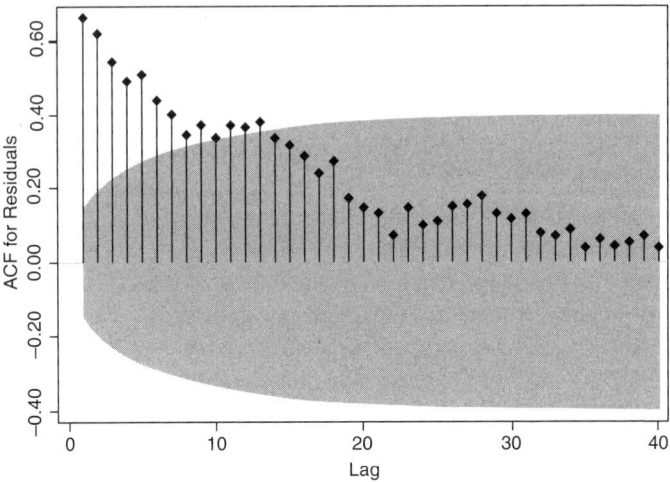

Bartlett's Formula for MA(q) 95% Confidence Bands

(b) Partial Autocorrelation Function

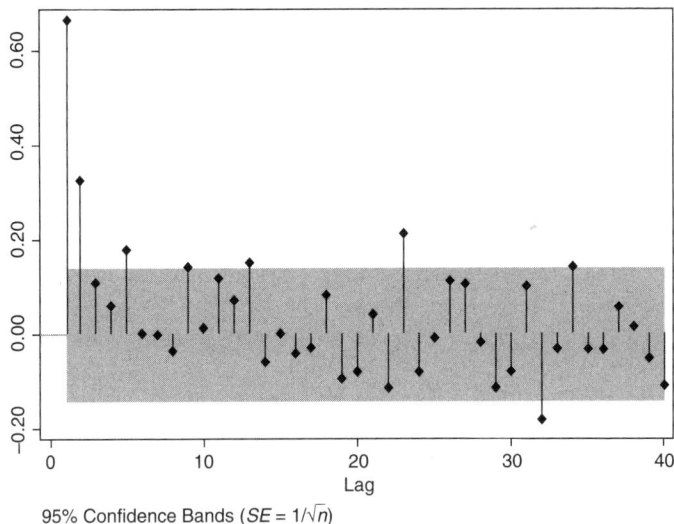

95% Confidence Bands ($SE = 1/\sqrt{n}$)

NOTE: ACF = autocorrelation function, MA = moving average, PACF = partial autocorrelation
function, SE = standard error.

Table 5.9 ln(KSI) in U.K. Traffic Accidents, January 1969 to February 1983

ln(KSI)	Coefficient	Standard Error	t Statistic	P Value
January	−0.24	0.019	−12.48	<0.001
February	−0.34	0.021	−16.34	<0.001
March	−0.31	0.023	−13.64	<0.001
April	−0.39	0.033	−11.79	<0.001
May	−0.30	0.025	−12.11	<0.001
June	−0.33	0.024	−13.53	<0.001
July	−0.28	0.028	−10.1	<0.001
August	−0.27	0.023	−11.57	<0.001
September	−0.25	0.023	−11.11	<0.001
October	−0.18	0.023	−7.68	<0.001
November	−0.061	0.021	−2.86	0.004
Petrol—ln(£)	−0.33	0.13	−2.57	0.01
Constant	6.93	0.29	23.74	<0.001
AR(1)	0.92	0.045	20.29	<0.001
MA(1)	−0.64	0.093	−6.89	<0.001

NOTE: Log likelihood = 211.317, T = 169; T = number of time points, AR = autoregressive, MA = moving average, KSI = killed or seriously injured.

After this time point, these are out-of-sample forecasts. We next plot the predicted values and forecasts and compare them to the raw ln(KSI) data (Figure 5.12).

We can see the consequence of not accounting for the seatbelt law after time point 169. This also gives us a nice visual representation of the estimated effect of the seatbelt law. At time point 170, the forecasted value of ln(KSI) clearly deviates from the observed value. This deviation appears to remain to the end of the observed time series. It would appear that a level change intervention occurring at time point 170 is appropriate. Therefore, we reestimate our model for the full period for which we have data and include the seatbelt law intervention as a step function (Table 5.10).

We next apply the Q test to the residuals from the model. The Portmanteau (Q) statistic is 54.72 and is chi-squared distributed with 40 degrees of freedom. The corresponding P value is 0.060. The Portmanteau

Figure 5.12   Predicted and Forecast ln(KSI)

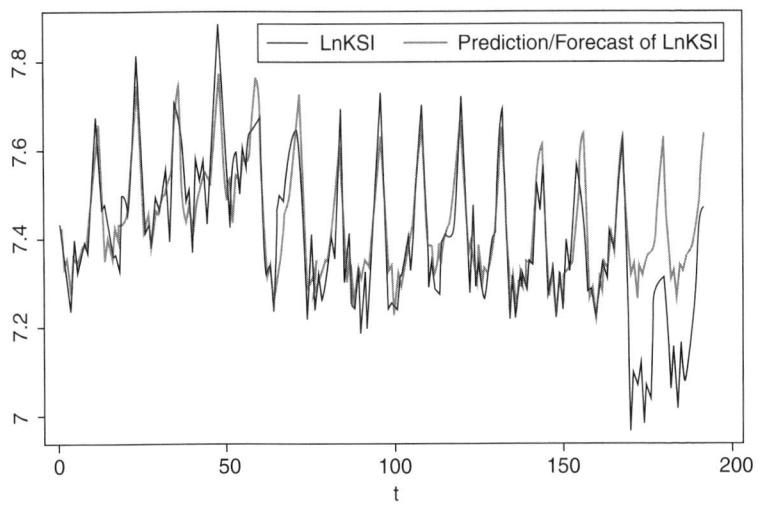

NOTE: KSI = killed or seriously injured.

Table 5.10   ln(KSI) in U.K. Traffic Accidents, January 1969 to December 1984

ln(KSI)	Coefficient	Standard Error	t Statistic	P Value
January	−0.23	0.018	−12.93	<0.001
February	−0.35	0.020	−17.44	<0.001
March	−0.31	0.022	−13.99	<0.001
April	−0.38	0.030	−12.95	<0.001
May	−0.30	0.023	−12.67	<0.001

Table 5.10

ln(KSI)	Coefficient	Standard Error	t Statistic	P Value
June	−0.33	0.022	−14.92	<0.001
July	−0.28	0.026	−10.88	<0.001
August	−0.27	0.022	−12.39	<0.001
September	−0.24	0.020	−11.71	<0.001
October	−0.16	0.021	−7.71	<0.001
November	−0.055	0.020	−2.75	0.006
Petrol—ln(£)	−0.30	0.12	−2.46	0.014
Seatbelt law—step function	−0.22	0.041	−5.32	<0.001
Constant	7.01	0.28	25.43	<0.001
AR(1)	0.92	0.041	22.31	<0.001
MA(1)	−0.66	0.087	−7.55	<0.001

NOTE: Log likelihood = 240.945, T = 192; T = number of time points, AR = autoregressive, MA = moving average, KSI = killed or seriously injured.

statistic indicates that we cannot reject the null hypothesis of the errors being a white noise process. Note that including additional autoregressive and/or moving average processes would produce a Portmanteau statistic with a larger P value, but following the principle of parsimony, we use the model with the fewest such components that produces residuals for which we cannot reject the null of a white noise process at our chosen significance level.

The estimated coefficient for the seatbelt law step function is −0.22, indicating that the introduction of the new seatbelt law produced a 22% reduction in KSIs. As this is an ARMA model, it is also the long-run effect of the seatbelt law.

What if we had used a pulse function for the seatbelt law variable? Table 5.11 displays the results of the estimates from such a model. The immediate effect of the seatbelt law is −0.21. This is of the same magnitude as before. However, because the seatbelt law is a pulse function, it drops to 0 after time point 170. The consequence of including a pulse function is to model an additive outlier intervention. This means that the effects of the seatbelt law change disappear after $T = 170$. Given the effect of the seatbelt law plotted in Figure 5.12, this seems inappropriate. This brings us to transfer functions.

Table 5.11 ln(KSI) in U.K. Traffic Accidents, January 1969 to December 1984

ln(KSI)	Coefficient	Standard Error	t Statistic	P Value
January	−0.23	0.019	−12.39	<0.001
February	−0.34	0.020	−17.06	<0.001
March	−0.31	0.021	−15.28	<0.001
April	−0.39	0.030	−13.02	<0.001
May	−0.30	0.025	−12.01	<0.001
June	−0.34	0.023	−14.74	<0.001
July	−0.28	0.027	−10.6	<0.001
August	−0.28	0.023	−11.99	<0.001
September	−0.24	0.021	−11.31	<0.001
October	−0.16	0.022	−7.51	<0.001
November	−0.056	0.019	−3.01	0.003
Petrol—ln(£)	−0.32	0.16	−1.96	0.05
Seatbelt law—pulse function	−0.21	0.050	−4.23	<0.001
Constant	6.93	0.37	18.74	<0.001
AR(1)	0.94	0.034	27.79	<0.001
MA(1)	−0.55	0.079	−6.98	<0.001

NOTE: Log likelihood = 234.871, $T = 192$; T = number of time points, AR = autoregressive, MA = moving average, KSI = killed or seriously injured.

Transfer Functions

In the last example, we assumed a lag structure for the independent variable, "petrol." This lag structure was based on theoretical expectations. In the social sciences, we generally have strong theories to suggest when an independent variable will have its effect on our dependent variable. However, there is sometimes some ambiguity about whether an independent variable has an effect that is immediate or has some degree of lag.

If an independent variable x_t is white noise, as in Equation 5.4.2, it is rather straightforward to examine the cross-correlations between y_t and x_t to determine the lag structure of x_t for the model of y_t.

$$y_t = \alpha_{0,1} + \alpha_{1,1} y_{t-1} + \beta_{1,1} x_t + \varepsilon_t, \tag{5.4.1}$$

$$x_t = \alpha_{0,2} + v_t. \tag{5.4.2}$$

The cross-correlations are the correlations between y_t and the lags of x_t: x_{t-1}, x_{t-2}, ... A plot of these cross-correlations is called a cross-correlogram. The use of a cross-correlogram to determine the lag structure in this manner makes the important assumption that there is no feedback from y_t to x_t. This means that our data-generating process for x_t (Equation 5.4.2) does not contain a current or lagged value of y_t. If such an assumption is not feasible for the contemporaneous values of x_t, we need to consider more advanced multivariate time series approaches, such as vector autoregression. This is an advanced topic, which receives excellent treatment in Brandt and Williams (2006) and Enders (2004).

As an example of the use of a cross-correlogram to determine the lag structure of an independent variable, consider a model of media tone toward the governing party as a function of errors in published vote intention polls. The data are derived from media coverage of the 2006 Canadian federal election campaign. The media content analysis involves a sample of mainstream newspaper, radio, television, and Internet coverage of the election during the campaign period.[3] Codes for media tone—*positive*, *negative*, or *neutral*—were assigned for all major parties. A party received a −1, 0, or +1 tone score each time it appeared in a story. Net tone, determined by the proportion of positive stories minus the proportion of negative stories, indicates the relative weight of positive news over negative news for a given party over all stories on a given campaign day. For this example, we use the net tone of media coverage for the governing Liberal party.

---

[3] For full methodological details of the content analysis, see Blake (2010).

The poll error (in percentage points) is based on the average of the differences between the poll results published each day of the campaign and estimates of true vote intention for the period the polls were in the field (Pickup, Andrew, Cutler, & Matthews, 2014). Again, we use the estimated poll error for the governing Liberal party. It is the change in this estimate of poll error that is used as the regressor. Polling error is random and (theoretically) independent from one poll to the next, and so it should be white noise, as in Equation 5.4.2, as should its first difference. The cross-correlogram is presented in Table 5.12.

The cross-correlogram indicates cross-correlations at the zero and first lags of poll error. On this basis, media tone is regressed on these lags of poll error. The results are presented in Table 5.13. The significant and positive coefficients for poll error and its lag suggest that errors that overstate the popularity of the governing Liberal party tend to produce positive news coverage the day the poll is published and the following day. A change of

Table 5.12 Cross-Correlogram: Media Tone and Poll Error

Lag	Cross-Correlations
0	0.33
1	0.16
2	−0.057
3	0.0057
4	−0.087
5	0.025

Table 5.13 Media Tone and Poll Error

Tone	Coefficient	Standard Error	t Statistic	P Value
D. Poll error	0.026	0.010	2.6	0.014
LD. Poll error	0.021	0.0088	2.42	0.021
Constant	−0.14	0.010	−13.94	<0.001

NOTE: $R^2 = 0.19$, $T = 38$; T = number of time points, L = lag of variable, D = difference of variable.

1 percentage point in the error in favor of the Liberal party results in a positive shift of 2 to 2.5 percentage points in net tone of the media coverage of the Liberal party each day.

If x_t is not a white noise process, the appropriate lag structure for x_t in a model of y_t is more difficult to determine. To see why this might be the case, let the data-generating process for y_t and x_t be

$$y_t = \alpha_{0,1} + \alpha_{1,1} y_{t-1} + \beta_{1,1} x_t + \varepsilon_t,$$ (5.4.3)

$$x_t = \alpha_{0,2} + \alpha_{1,2} x_{t-1} + v_t.$$ (5.4.4)

If we plug Equation 5.4.4 into Equation 5.4.3, we get the following:

$$y_t = \alpha_{0,1} + \alpha_{1,1} y_{t-1} + \beta_{1,1}\left(\alpha_{0,2} + \alpha_{1,2} x_{t-1} + v_t\right) + \varepsilon_t,$$

$$y_t = (\alpha_{0,1} + \beta_{1,1}\alpha_{0,2}) + \alpha_{1,1} y_{t-1} + \beta_{1,1}\alpha_{1,2} x_{t-1} + \beta_{1,1} v_t + \varepsilon_t.$$ (5.4.5)

It might seem that the appropriate model for y_t includes a lag of x_t. There is nothing wrong with the estimation of such a model, but it would be incorrect to interpret this to mean that x_t causes y_t with a lag. If Equations 5.4.3 and 5.4.4 do represent the data-generating process, then the lag of x_t only has an effect on y_t through its effect on x_t. When the dynamic structure of y_t and x_t becomes more complicated, there is an increasing number of possible models for y_t and no clear-cut method for distinguishing between them.

In response to this challenge, Box and Jenkins (1976) proposed an inductive method for selecting the optimal lag structure for the purposes of forecasting. This requires the transfer function representation of the ARMA model (Harvey, 1993, subsection 5.8). Consider the two-component representation of the ARMA model with a single exogenous regressor, x_t:

$$y_t = \beta_0 + f_t + \mu_t,$$

$$\mu_t = \sum_{i=1}^{p} \alpha_i \mu_{t-i} + \sum_{j=1}^{q} \phi_j \varepsilon_{t-j} + \varepsilon_t.$$ (5.4.6)

The transfer function f_t is a function of x_t and determines how movements in the independent variable are translated into movements in the dependent variable. For example, a change in x_t might simply have an immediate and permanent effect, in which case f_t would simply be

$$f_t = \beta_1 x_t.$$

Of course, a change in x_t might have a more complicated effect, such as an immediate short-run effect that builds into a permanent long-run effect,

$$f_t = \varrho f_{t-1} + \beta_1 x_t,$$

assuming that $|\varrho| < 1$. Unfortunately, the transfer function representation does not lend itself to a systematic approach when there is more than one explanatory variable (Harvey, 1990, subsection 7.5). Furthermore, the Box-Jenkins approach is designed to optimize the value of the model for forecasting and not for testing hypotheses. Therefore, while it is important to consider alternative lag structures for the independent variables, the possibilities considered are best driven by theoretical considerations. In the context of intervention analysis, where the lag structure of an intervention is often less clear and may not be suggested to us by theory, the transfer function is a useful concept. Therefore, we examine the idea of a transfer function in that context.

Let the transfer function f_t have the following form:

$$f_t = \varrho f_{t-1} + \beta_1 x_t. \tag{5.4.7}$$

The ϱ parameter can be assumed to be 1, assumed to be 0, or estimated assuming that $|\varrho| < 1$. The *indicator variable* x_t could be either a step function or a pulse function and indicates a disturbance at a particular time $t = d$. If f_t is deterministic, $x_t = 0$ for $t < d$, $x_t = 1$ for $t = d$, and depending on whether x_t is a step or pulse $x_t = 1$ or $x_t = 0$ for $t > d$. If f_t is stochastic, $x_t = 0$ for $t < d$, $x_t \sim N(0, \sigma_v^2)$ for $t = d$ (Tsay, 1988), and depending on whether x_t is a step or pulse $x_t \sim N(0, \sigma_v^2)$ or $x_t = 0$ for $t > d$ (Box & Tiao, 1975).

For example, if we wish to model a deterministic step (aka level) change intervention, x_t is a deterministic step function and ϱ is assumed to be 0:

$$f_t = \beta_1 x_t. \tag{5.4.8}$$

Alternatively, ϱ could be estimated, and f_t would model a deterministic dynamic step change:

$$f_t = \varrho f_{t-1} + \beta_1 x_t. \tag{5.4.9}$$

If, instead, ϱ is assumed to be 1, f_t produces a deterministic ramp response:

$$f_t = f_{t-1} + \beta_1 x_t. \tag{5.4.10}$$

A ramp response builds over time, increasing by β_1 each time point. If x_t in Equation 5.4.9 is a pulse function, instead of a step function,

f_t produces an effect that decays to zero at a rate determined by ϱ. This last intervention is an example of a transient change intervention. To produce more complicated interventions, multiple transfer functions can be combined. If we wish the intervention to produce an effect that decays to a nonzero value, we can include two transfer functions in Equation 5.4.6:

$$f_{1,t} = \varrho f_{1,t-1} + \beta_1 x_t.$$

$$f_{2,t} = f_{2,t-1} + \beta_2 x_t. \tag{5.4.11}$$

Assuming that $|\varrho| < 1$ and with x_t defined as a pulse function, the initial effect will have a magnitude of $\beta_1 + \beta_2$, and this will decay to a magnitude of β_2. While each of these interventions has been deterministic, they can be modelled as stochastic by using the stochastic form of x_t.

If we combine Equation 5.4.6 with Equation 5.4.11 or any other transfer function, we have what is called a state-space model. This is an advanced topic, and the interested reader is referred to Commandeur and Koopman (2007).[4] Such a model is called a structural model as the included terms and parameters reflect the theorized structure of the data-generating process.

$$y_t = \beta_0 + f_{1,t} + f_{2,t} + \mu_t,$$

$$\mu_t = \sum_{i=1}^{p} \alpha_i \mu_{t-i} + \sum_{j=1}^{q} \phi_j \varepsilon_{t-j} + \varepsilon_t.$$

$$f_{1,t} = \varrho f_{1,t-1} + \beta_1 x_t.$$

$$f_{2,t} = f_{2,t-1} + \beta_2 x_t. \tag{5.4.12}$$

If we wish to estimate such a model as an ARMA or ADL model, we would need to transform it. If our model contains only a single autoregressive term and the transfer function from Equation 5.4.9, an appropriate transformation would be as follows:

$$y_t = \beta_0^* + \alpha_1 y_{t-1} + \beta_1^* x_t + \varepsilon_t,$$

---

[4] Some examples of the use of state-space models in the social sciences are Martin and Quinn (2002), Pickup and Johnston (2008), Armstrong (2008), Jackman (2005), Beck (1989), McAvoy (1998), Brandt and Williams (2001), and Kellstedt, McAvoy, and Stimson (1996).

where $\beta_0^* \equiv (1 - \alpha_1)\beta_0$ and $\beta_1^* \equiv \frac{\alpha_1}{\varrho}\beta_1$. The transformation is straightforward but involved and so is not included here. Such a model could be estimated by OLS, but note that we do not estimate the structural parameters.

In the last section of this chapter, we extend our discussion of ARCH models from Chapter 4.

5.5 Generalized Autoregressive Conditional Heteroskedasticity (GARCH) Models

In Chapter 4, we described an LDV(1) model with an ARCH process as follows:

$$y_t = \alpha_0 + \alpha_1 y_{t-1} + \beta_1 x_t + \varepsilon_t,$$

$$\varepsilon_t = \sqrt{h_t}\, v_t. \tag{5.5.1}$$

where v_t is a white noise process with a zero mean and unit variance.

$$E(v_t) = 0.$$

$$E\left(v_t v_{t-s}\right) = \begin{cases} 1 \text{ for } s = 0 \\ 0 \text{ otherwise} \end{cases}.$$

The unconditional variance of ε_t is constant, and the conditional variance of ε_t is h_t—a function of past ε_t^2 (Hamilton, 1994). Specifically,

$$h_t = \zeta + \phi_1 \varepsilon_{t-1}^2 + \phi_2 \varepsilon_{t-2}^2 + \cdots + \phi_m \varepsilon_{t-m}^2. \tag{5.5.2}$$

A generalization of the ARCH model (GARCH) is to include lags of the conditional variance in Equation 5.5.2 (Bollerslev, 1986):

$$h_t = \zeta + \phi_1 \varepsilon_{t-1}^2 + \phi_2 \varepsilon_{t-2}^2 + \cdots + \phi_m \varepsilon_{t-m}^2 + \delta_1 h_{t-1} + \delta_2 h_{t-2} + \cdots + \delta_r h_{t-r}. \tag{5.5.3}$$

From this, the squared errors can be represented as follows (Hamilton, 1994):

$$\varepsilon_t^2 = \zeta + \left(\delta_1 + \phi_1\right)\varepsilon_{t-1}^2 + \left(\delta_2 + \phi_2\right)\varepsilon_{t-2}^2 + \cdots + \left(\delta_p + \phi_p\right)\varepsilon_{t-p}^2$$

$$+ \omega_t - \delta_1 \omega_{t-1} - \delta_2 \omega_{t-2} - \cdots - \delta_r \omega_{t-r}, \tag{5.5.4}$$

where p is the larger of m and r and $\omega_t = \varepsilon_t^2 - h_t$. We denote the order of the GARCH as GARCH(r, m). In this representation, ε_t^2 follows an ARMA(p, r)

process (Bollerslev, 1986). This fact is useful for determining the order of the GARCH model to use, as we will see momentarily. One potential advantage of this extension of the ARCH model is that a higher-order ARCH process can often be represented by a lower-order GARCH process, thereby providing gains in efficiency.

As in the ARCH process, the conditional and unconditional means of ε_t are zero, and the unconditional variance of ε_t is a constant. In particular,

$$E(\varepsilon_t^2) = E(h_t v_t^2) = E(h_t)$$

$$= \frac{\zeta}{1 - (\delta_1 + \phi_1) - (\delta_2 + \phi_2) - \cdots - (\delta_p + \phi_p)}. \tag{5.5.5}$$

The covariance stationarity requirement for this process is that

$$(\delta_1 + \phi_1) + (\delta_2 + \phi_2) + \cdots + (\delta_p + \phi_p) < 1. \tag{5.5.6}$$

Just as an ARCH or GARCH process can be included within an LDV model, they can also be included within an ARMA model. Using the two-component representation of the ARMA model (Equation 5.3.1), we can include a GARCH(1,1) process in an ARMA(p,q) model as follows:

$$y_t = \beta_0 + \sum_{j=1}^{k} \beta_j x_j + \mu_t,$$

$$\mu_t = \sum_{i=1}^{p} \alpha_i \mu_{t-i} + \sum_{j=1}^{q} \phi_j \varepsilon_{t-j} + \varepsilon_t.$$

$$\varepsilon_t^2 = \zeta + (\delta_1 + \phi_1) \varepsilon_{t-1}^2 + \omega_t + \delta_1 \omega_{t-1}. \tag{5.5.7}$$

To demonstrate the specification, estimation, and testing of a GARCH model, we return to the German economic approval model we used in Chapter 4 to demonstrate the ARCH model. After estimating the ADL(1,1) model, we might have produced the ACF and PACF for model residuals, as shown in Figure 5.13.

The ACF and PACF suggest that we might have wanted to include either an AR(1) or an MA(1) term in our model, but this model is already autoregressive by including a lagged dependent variable, so we proceed with the ADL(1,1) MA(1) model. The estimates of this model are presented in Table 5.14. This specification produces the desired white noise errors, according to the Portmanteau Q statistic. Next, we calculate the

Figure 5.13 Autocorrelation and Partial Autocorrelation Functions for German Economic Approval

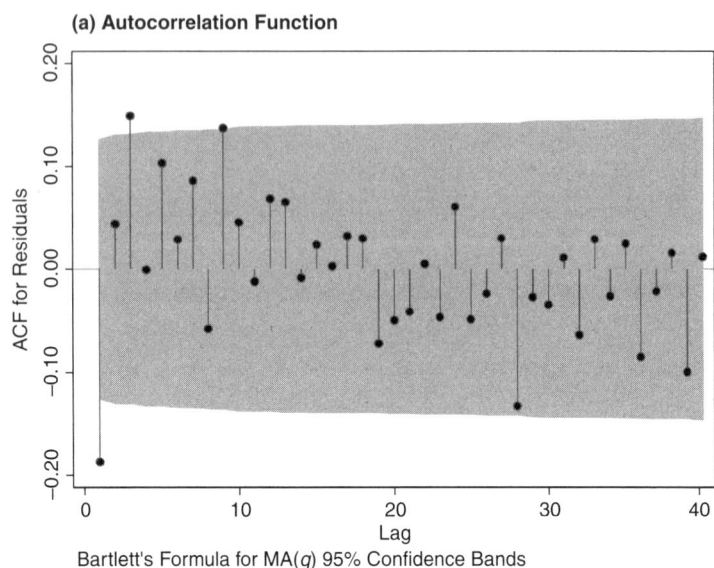

(a) Autocorrelation Function

Bartlett's Formula for MA(*q*) 95% Confidence Bands

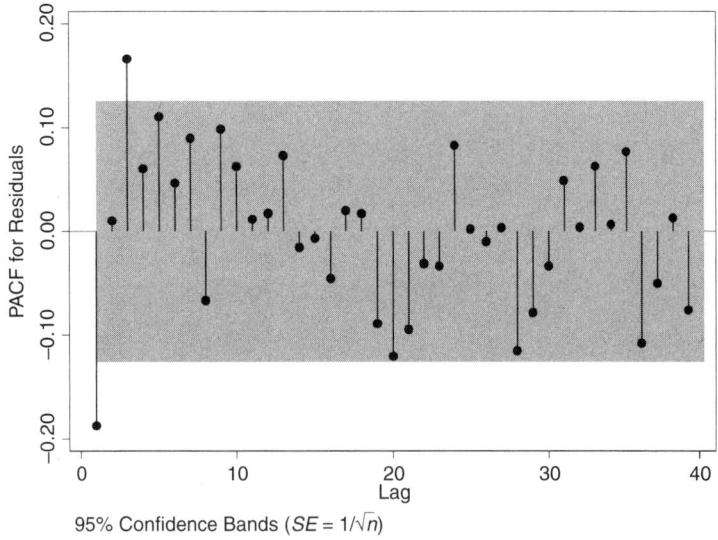

(b) Partial Autocorrelation Function

95% Confidence Bands ($SE = 1/\sqrt{n}$)

NOTE: ACF = autocorrelation function, MA = moving average, PACF = partial autocorrelation function, SE = standard error.

Table 5.14 ADL(1,1) MA(1) German Approval Model With GARCH(1,1) Process

Approval	ADL(1,1) MA(1)				ADL(1,1) MA(1) GARCH(1,1)			
	Coefficient	Standard Error	z Statistic	P Value	Coefficient	Standard Error	z Statistic	P Value
L1. Approval	0.90	0.039	23.06	<0.001	0.87	0.044	19.55	<0.001
GDP	0.41	0.27	1.54	0.123	0.52	0.26	1.99	0.046
Unemployment	−0.36	2.32	−0.16	0.876	−0.25	3.17	−0.08	0.936
Inflation	0.13	0.95	0.14	0.888	0.42	0.76	0.56	0.578
L1. GDP	−0.36	0.27	−1.36	0.173	−0.46	0.25	−1.83	0.067
L1. Unemployment	0.11	2.33	0.05	0.963	−0.15	3.17	−0.05	0.962
L1.Inflation	−0.40	0.94	−0.43	0.67	−0.87	0.75	−1.15	0.249
Constant	2.20	1.15	1.91	0.056	3.76	1.23	3.07	0.002
MA(1)	−0.33	0.055	−6	<0.001	−0.18	0.098	−1.85	0.064
ARCH(1)	—	—	—	—	0.17	0.056	3.11	0.002
GARCH(1)	—	—	—	—	0.62	0.11	5.49	<0.001
Constant	—	—	—	—	2.94	1.28	2.30	0.021
	Log likelihood = −669.7142				Log likelihood = −653.3762			

NOTE: $N = 242$; P = probability, ADL = autoregressive distributed lag, ARCH = autoregressive conditional heteroskedasticity, GARCH = generalized autoregressive conditional heteroskedasticity, MA = moving average, L1 = first lag.

autocorrelation and partial autocorrelation functions for the squared residuals of the model to help us determine the specification of the GARCH process. These are shown in Figure 5.14.

The squared residuals of a GARCH(r,m) follow an ARMA(p,r) process with $p = \max\{m,r\}$. The ACF and PACF of the squared residuals indicate that there is an MA(1) process. Accordingly, we can deduce that $r = 1$. Since $p = \max\{m,r\}$ and $r = 1$, we would expect p to be at least 1. This means that we would expect to see at least an AR(1) component in the squared residuals. This is not evident in Figure 5.14, but it may just be too small to be observed, and/or the MA(1) process in the squared residuals might be obscuring evidence of it.

The value of m is more ambiguous in this case. If we assume that $m = 0$, we have the ARCH(1) process we used in our model in Chapter 4. However, there is nothing in Figure 5.14 that rules out the possibility that $m = 1$ and there is no evidence it is any larger. Therefore, we will include a GARCH(1,1) process and compare (Table 5.14).

Testing the residuals, the Q statistic is 30.79 with a P value of 0.85. We cannot reject the null of a white noise process. The ACF and PACF for the residuals (Figure 5.15) also indicate the residuals are white noise. We next test the adequacy of the specification of the GARCH process by examining the standardized residuals. These should also be a white noise process. We calculate standardized residuals by dividing the estimated errors by the square root of the estimated conditional variance of those errors: $\hat{s}_t = \hat{\varepsilon}_t / \sqrt{\hat{h}_t}$. The ACF and PACF for the standardized residuals are shown in Figure 5.16.

There is nothing in the ACF and PACF for the standardized residuals to suggest that they are anything but a white noise process. Furthermore, the Q statistic is 34.08 with a P value of 0.73. If there had been a pattern that suggested dynamics other than white noise, we could use the ACF and PACF for the squared residuals to inform the respecification of the GARCH process.

Having determined that the specification of the GARCH process is adequate, we can compare the results from the ADL(1,1) MA(1) GARCH(1,1) and the ADL(1,1) MA(1) (Table 5.14). Neither of the growth in GDP coefficients are statistically significant (at the 0.5 significance level) in the ADL(1,1) MA(1) model. This is also true for the estimated long-run effect of growth. (This is calculated in the same manner as for any ADL model results.) In the same model, neither inflation nor unemployment has a statistically significant short-run or long-run effect. In the ADL(1,1) MA(1) GARCH(1,1) model, the contemporaneous effect of growth is statistically significant, although the estimated long-run effect is not. Inflation and unemployment do have statistically significant long-run effects of -3.46 and -3.08, respectively. The inclusion of the GARCH(1,1) process in the model reveals that there is a short-lived effect of growth in GDP on government approval. It also reveals the long-run effects of inflation and unemployment.

Figure 5.14 Autocorrelation and Partial Autocorrelation Functions for the Squared Residuals

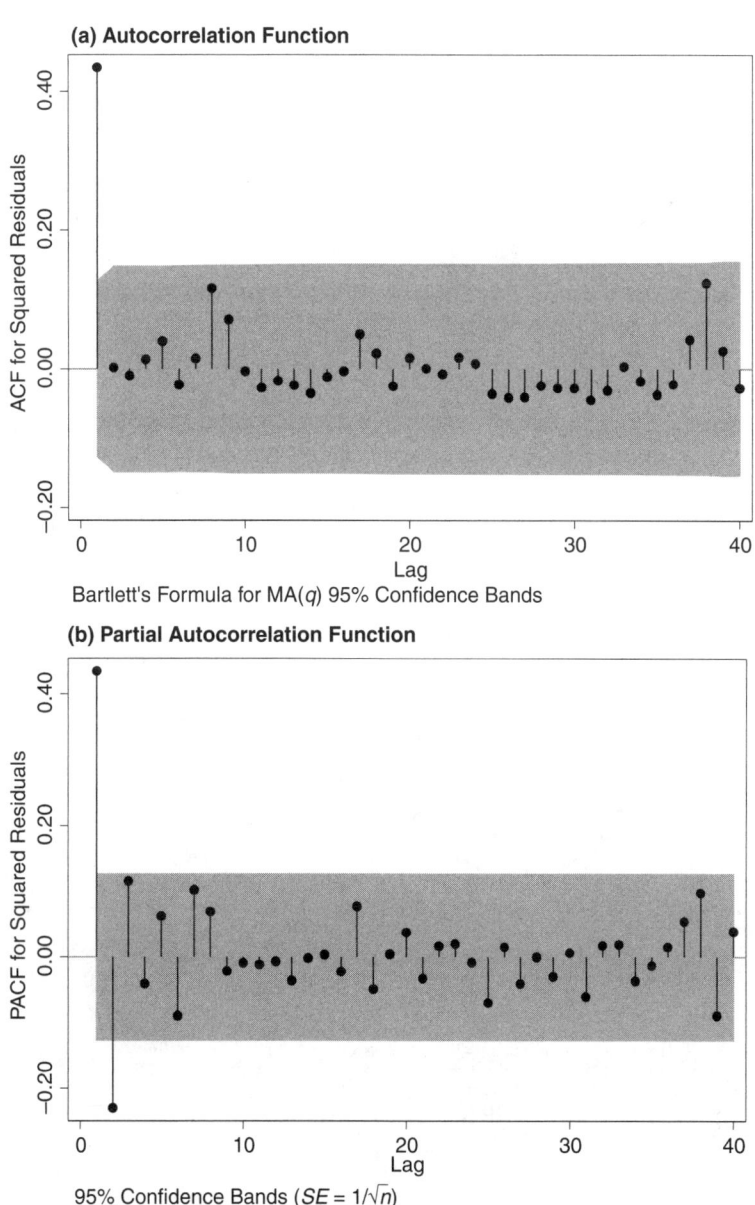

NOTE: ACF = autocorrelation function, MA = moving average, PACF = partial autocorrelation function, SE = standard error.

Figure 5.15 Autocorrelation and Partial Autocorrelation Functions for
GARCH Residuals

(a) Autocorrelation Function

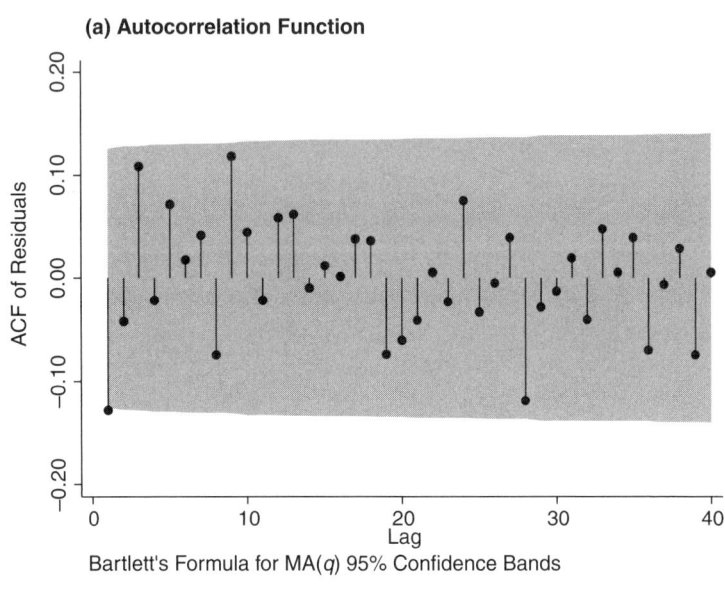

Bartlett's Formula for MA(*q*) 95% Confidence Bands

(b) Partial Autocorrelation Function

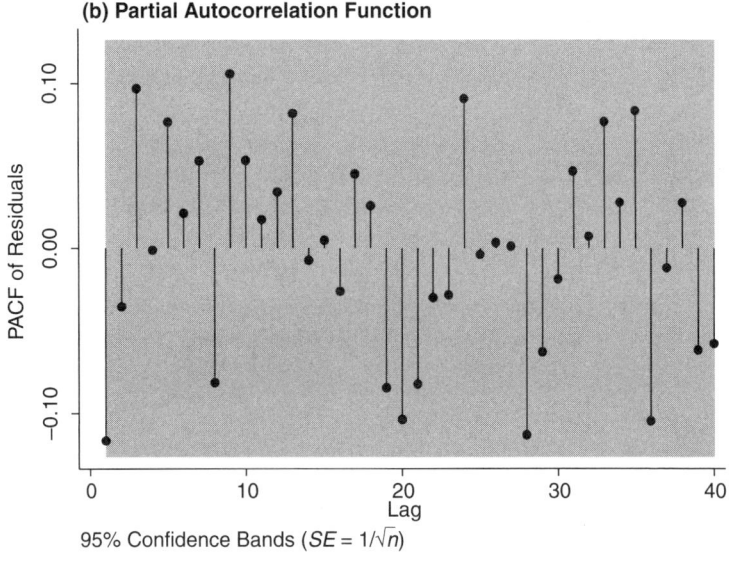

95% Confidence Bands ($SE = 1/\sqrt{n}$)

NOTE: ACF = autocorrelation function, MA = moving average, PACF = partial autocorrelation function, SE = standard error.

Figure 5.16   Autocorrelation and Partial Autocorrelation Functions for the Standardized Residuals

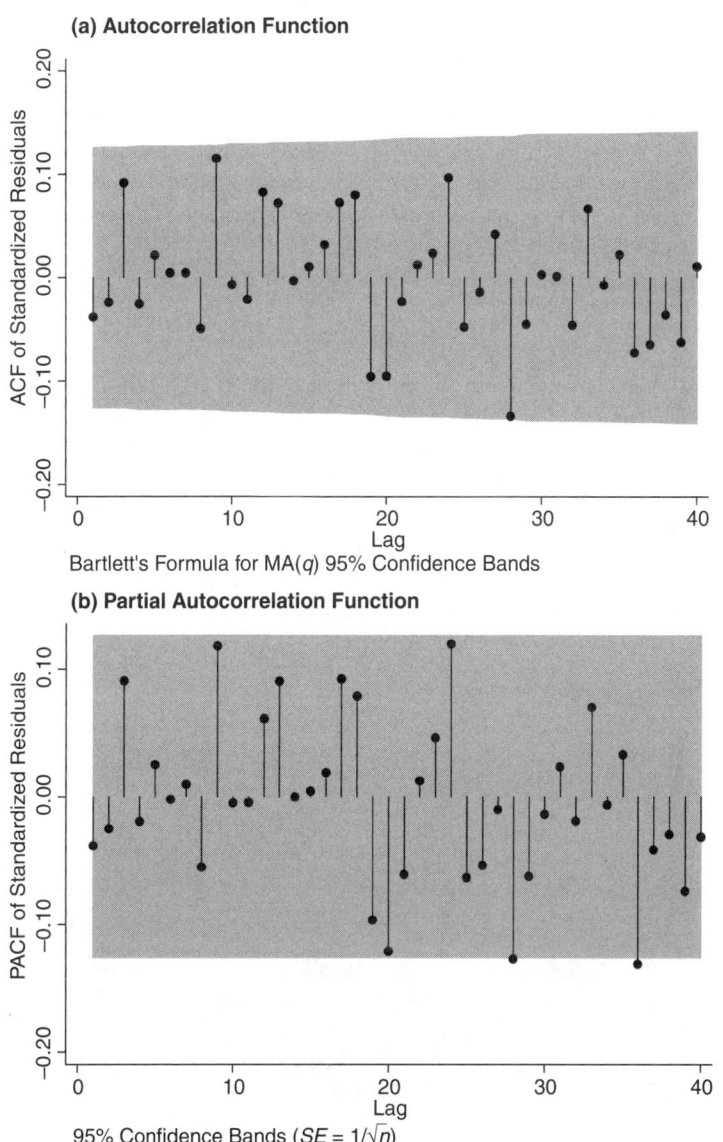

NOTE: ACF = autocorrelation function, MA = moving average, PACF = partial autocorrelation function, SE = standard error.

We now compare the fit of the ADL(1,1) MA(1) GARCH(1,1) with the fit of the ADL(1,1) MA(1) ARCH(1). The second is nested within the first, so we can use the LR test. The test statistic is chi-squared distributed with 1 degree of freedom. It is 3.27, with a corresponding P value of 0.0704. We cannot reject the null hypothesis that the ADL(1,1) MA(1) GARCH(1,1) fits no better than the ADL(1,1) MA(1) ARCH(1) at the 0.05 significance level, although we could at the 0.1 significance level. If we tested the standardized residuals from the ADL(1,1) MA(1) ARCH(1) model the tests would suggest that the standardized residuals are a white noise process and that the ARCH(1) process is an adequate specification. The evidence suggests that either model is adequate.

Summary

In this chapter, you have been introduced to the Box-Jenkins approach to model selection. This approach is distinct from the general-to-specific approach discussed in Chapter 4. The Box-Jenkins approach places an emphasis on parsimony and building up a minimal model that meets the estimation assumptions for the data. In contrast, the general-to-specific approach starts with the most general model and only reduces the model through the placement of restrictions if they can be justified by both theory and data. The GARCH model discussed in the last section of this chapter can be incorporated into either approach.

CHAPTER 6: MODELS FOR INTEGRATED AND COINTEGRATED DATA

Chapter 6 extends the discussion of autoregressive moving average (ARMA) models, begun in Chapter 5, to the autoregressive integrated moving average (ARIMA) model and error correction model (ECM) by discussing the concept of differencing integrated data. The reader is also introduced to seasonal and fractional differencing. To that end, the chapter begins with an overview of identifying unit root processes and how one may distinguish these from other forms of trending. Differencing data and the differenced data model are then discussed, and this is used to motivate the ARIMA model and ECM.

6.1 Unit Root Processes and Differencing Data

A well-known example of a unit root process is the random walk (Pearson, 1905). A random walk is a process in which the current value of the series is equal to the last value plus a white noise term:

$$y_t = y_{t-1} + \varepsilon_t. \tag{6.1.1}$$

With a random walk, $\text{Var}(y_{t+h}) = \text{Var}(y_t) + h\sigma_\varepsilon^2$, so it increases with time (h) and is not stationary. In fact, the variance is explosive. We say that a random walk is highly persistent since $E(y_{t+h}|y_t) = y_t$ for all $h \geq 1$, meaning that the series is not weakly dependent. The interpretation of this is that at time t_0 our best forecast of $y_{t=t_0+h}$ is $y_{t=t_0}$ (the current value), no matter how far in the future $t_0 + h$ might be. This is distinct from a stationary process, for which there is an equilibrium. In a stationary process, the equilibrium is our best forecast for the long-term future ($h \to \infty$).

Note that trending and persistence are different things. A series can be trending but not persistent (e.g., a linear, deterministic trend), or a series can be highly persistent without any trend (e.g., a random walk). A random walk with drift is an example of a data-generating process that is both highly persistent and trending:

$$y_t = y_{t-1} + \alpha_0 + \varepsilon_t. \tag{6.1.2}$$

In addition to the variance increasing with time, so does the mean:

$$E(y_{t+h} | y_t) = y_t + h\alpha_0. \tag{6.1.3}$$

The series increases by α_0 with each time point. A series can also be a random walk with trend:

$$y_t = y_{t-1} + \alpha_1 t + \varepsilon_t. \tag{6.1.4}$$

Note that although a random walk process without a drift or trend does not have a mean that increases (or decreases) with time, it is sometimes called a stochastic trend process.

A random walk is a special case of a unit root process. Unlike some nonstationary processes, unit root processes have the property that they can be transformed into a stationary process through differencing one (if they contain a single unit root) or more (if they contain multiple unit roots) times. These processes are called integrated of order d, and the degree of integration (d) is the number of times they must be differenced to become stationary. This is denoted as $I(d)$.

A random walk is an $I(1)$ process as it is made stationary—$I(0)$—by differencing once. We can easily see how a random walk or a random walk with drift can be made stationary through first differencing. Subtract y_{t-1} from both sides of Equation 6.1.2:

$$y_t - y_{t-1} = y_{t-1} - y_{t-1} + \alpha_0 + \varepsilon_t,$$

$$y_t - y_{t-1} = \Delta y_t = \alpha_0 + \varepsilon_t.$$

The resulting Δy_t is clearly an $I(0)$ process. It is a constant plus white noise.

Not correcting for the fact that a data series comes from an $I(1)$ data-generating process can lead to spurious regression problems (Granger & Newbold, 1974; Phillips, 1986). Consider running a simple regression of y_t on x_t where y_t and x_t have independent $I(1)$ data-generating processes. The usual ordinary least squares (OLS) t statistic will often indicate a statistically significant relationship when there is none.

One solution is to simply difference the $I(1)$ data series, producing $I(0)$ series. Generally speaking, a series from a data-generating process that does not have a fixed equilibrium—such as the random walk—can be transformed through differencing (sometimes more than once[1]) to produce a stationary series. However, it is important not to overdifference data.

If y_t has a stationary or even a trend-stationary data-generating process, first differencing is inappropriate. To see why, suppose that y_t has an

---

[1] First differencing a time series that has already been first differenced is called second differencing.

AR(1) (autoregressive process of order 1) with a trend (trend stationary) data-generating process:

$$y_t = \alpha_0 + \alpha_1 y_{t-1} + \beta_1 t + \varepsilon_t. \tag{6.1.5}$$

Then,

$$y_{t-1} = \alpha_0 + \alpha_1 y_{t-2} + \beta_1 (t-1) + \varepsilon_{t-1}. \tag{6.1.6}$$

From Equations 6.1.5 and 6.1.6, we get

$$y_t - y_{t-1} = \alpha_0 + \alpha_1 y_{t-1} + \beta_1 t + \varepsilon_t - (\alpha_0 + \alpha_1 y_{t-2} + \beta_1 (t-1) + \varepsilon_{t-1}),$$

$$\Delta y_t = \alpha_1 (y_{t-1} - y_{t-2}) + \beta_1 t - \beta_1 (t-1) + \varepsilon_t - \varepsilon_{t-1},$$

$$\Delta y_t = \alpha_1 (y_{t-1} - y_{t-2}) + \beta_1 + \varepsilon_t - \varepsilon_{t-1},$$

$$\Delta y_t = \alpha_1 \Delta y_{t-1} + \beta_1 + \varepsilon_t - \varepsilon_{t-1}. \tag{6.1.7}$$

We have eliminated the trend through differencing, but we have introduced an MA(1) random walk into the errors: $\varepsilon_t - \varepsilon_{t-1}$. This is an MA(1) process with $\phi = 1$. Maximum likelihood estimation can run into problems when estimating data from such a data-generating process, particularly with a small sample size (Enders, 2004, p. 167). This is referred to as "overdifferencing."

To know whether or not we need to first difference our data, we must know if we really have a unit root process, a unit root with drift process, a stationary process, or a trend-stationary process. However, this can be difficult. Figure 6.1 plots a realization from each of the following data-generating processes: random walk with drift, random walk with trend, trend stationary, and AR(1) trend stationary.

Random walk with drift	$y_t = y_{t-1} + 0.5 + \varepsilon_t.$
Random walk with trend	$y_t = y_{t-1} + 0.01t + \varepsilon_t.$
Trend stationary	$y_t = 0.4t + \varepsilon_t.$
AR(1) trend stationary	$y_t = 0.9 y_{t-1} + 0.01t + \varepsilon_t.$

Note that plots of these processes are visually very similar. It is often difficult to distinguish a random walk (with drift or trend) process from a trend-stationary process, but there are test procedures to assist us. Before turning to how we might go about doing this, let us discuss seasonal differencing.

168

Figure 6.1   Trending Series

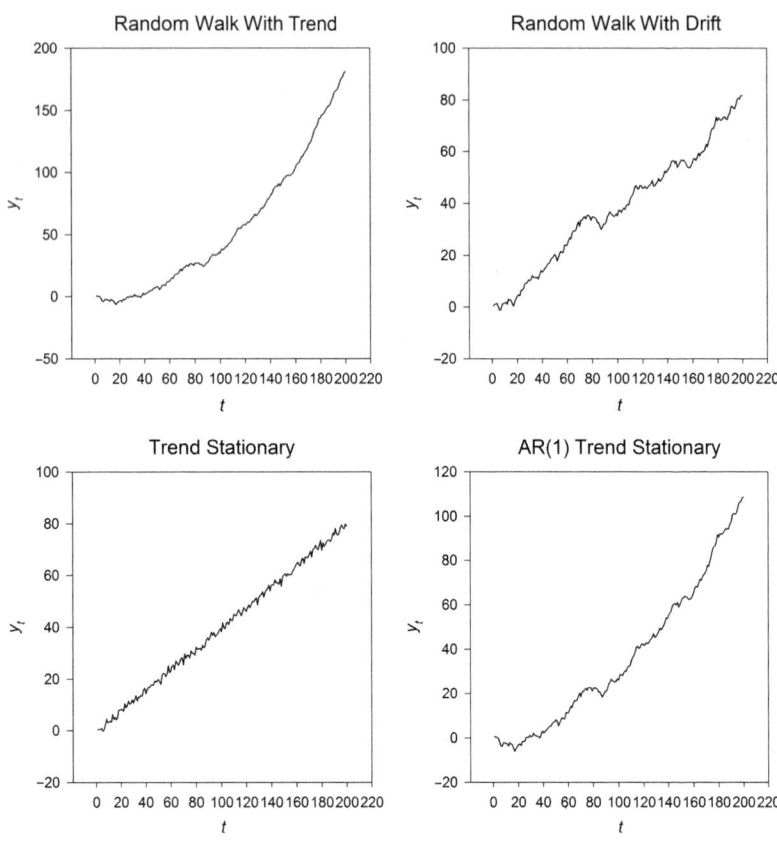

NOTE: AR = autoregressive.

In Chapter 3, we first examined the KSI (killed or seriously injured) data and noted that they had a distinct seasonal pattern. We accounted for this by including a series of monthly indicator variables in our analysis. An alternative solution is to seasonally difference the data. In this instance, we have monthly data, and the pattern repeats itself each year. Seasonal differencing in this case would be to take the 12th difference: $\Delta_{12}\, y_t = y_t - y_{t-12}$. After seasonal differencing, the data represent the change since the same month the previous year. Figure 6.2 presents plots of the original ln(KSI) data and the 12th difference. The seasonal pattern that is so evident in the original data is removed from the seasonally differenced data.

Figure 6.2   ln(KSI) and the 12th Difference

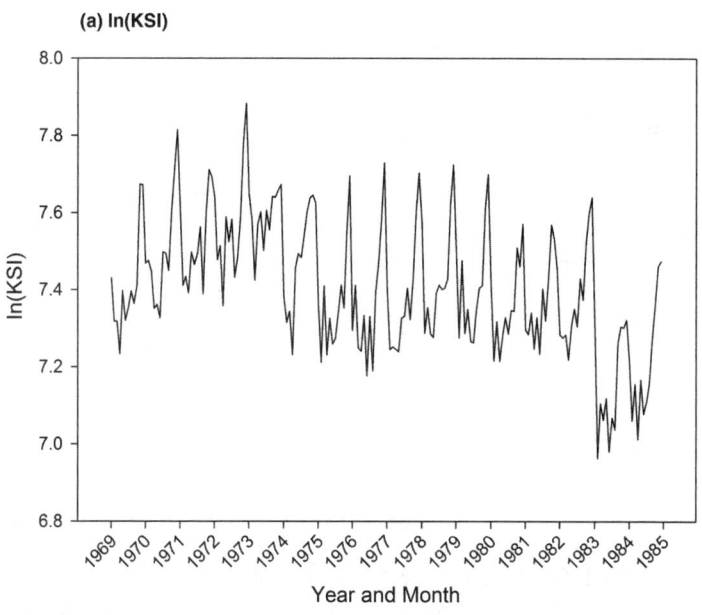

(a) ln(KSI)

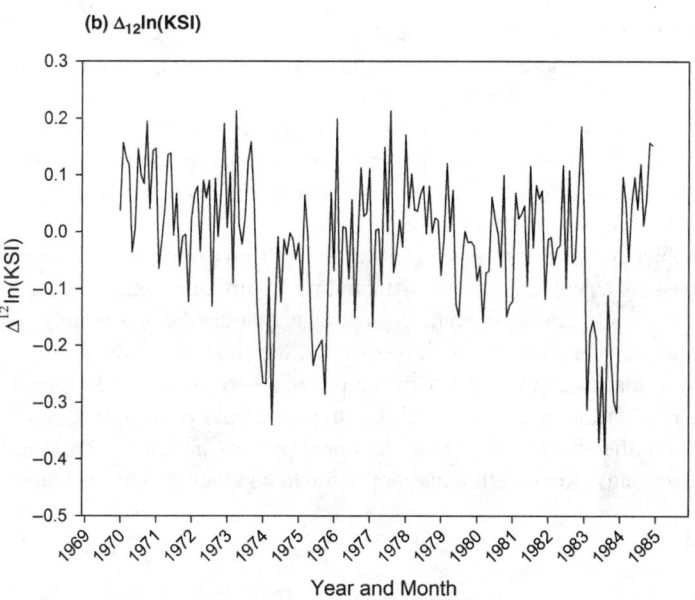

(b) Δ_{12}ln(KSI)

Data measured in other temporal units can also be seasonally differenced. For example, quarterly data can be seasonally differenced by taking the fourth difference. We will discuss the use of seasonally differenced data in ARIMA models in Section 6.3.

6.2 Testing for Unit Roots

We now consider tests that can help us distinguish between unit root (with or without drift or trend) and (trend-) stationary processes. Consider the following data-generating process:

$$y_t = \alpha_1 y_{t-1} + \varepsilon_t. \tag{6.2.1}$$

Subtract y_{t-1} from both sides of Equation 6.2.1, and define $\theta = \alpha_1 - 1$ to obtain

$$y_t - y_{t-1} = \alpha_1 y_{t-1} - y_{t-1} + \varepsilon_t,$$

$$\Delta y_t = (\alpha_1 - 1) y_{t-1} + \varepsilon_t,$$

$$= \theta y_{t-1} + \varepsilon_t. \tag{6.2.2}$$

Let $H_0 : \alpha_1 = 1$. That is, assume the null hypothesis that y_t is a unit root process. We can test that $\theta = 0$ by regressing Δy_t on y_{t-1}. Since $\theta = \alpha_1 - 1$, testing the null hypothesis that $\theta = 0$ is equivalent to testing the null hypothesis that $\alpha_1 = 1$. Rejecting the null $\theta = 0$ means rejecting $\alpha_1 = 1$. This means rejecting the null of a unit root data-generating process, with the alternative being an AR(1) process.

Unfortunately, a simple t test is inappropriate, since the data for y_{t-1} are from an I(1) data-generating process under the assumed null hypothesis. If it is a unit root process, then the assumptions necessary for OLS regression are not met, and we cannot use the t test.

A Dickey-Fuller test tests that a variable is from a unit root data-generating process by calculating the t statistic but using different critical values to test that $\theta = 0$.[2] As above, the null hypothesis is that the data-generating process for the variable contains a unit root: $\theta = 0$, which is to say, $\alpha_1 = 1$.

Note that Equation 6.2.1 does not contain a constant, and so the alternative is a stationary process with the unconditional expected value of 0. If we wish the alternative to be a stationary process including a constant, we can, instead, use the following regression to conduct the Dickey-Fuller test:

$$\Delta y_t = \alpha_0 + \theta y_{t-1} + \varepsilon_t. \tag{6.2.3}$$

---

[2] The test used is one-sided (left-tailed), as we are assuming α_1 is not greater than 1 and so θ is not greater than 0.

For the null hypothesis that y_t is a unit root process, we test that $\theta = \alpha_0 = 0$ (Hamilton, 1994).

To get a better understanding of the properties of the Dickey-Fuller test, let us run it on data that we generate randomly and where we specify the data-generating process (e.g., I(1) or AR(1)). By doing so, we know the correct test results. To begin, we generate a single realization of the following data-generating process: $y_t = y_{t-1} + \varepsilon_t$. This is done by randomly generating 300 independent values of ε_t from a standard normal distribution: NID(0, 1). We next define $y_1 = \varepsilon_1, y_2 = y_1 + \varepsilon_2, y_3 = y_2 + \varepsilon_3$, and so on. The resulting series $\{y_1, y_2, \ldots, y_{300}\}$ is a single realization of a unit root process with $T = 300$.

We apply the Dickey-Fuller test for a unit root process to this realization. The test statistic is -1.57. The 5% critical value is -2.88. As we would hope, we are unable to reject the null hypothesis of a unit root process. If we did this many times over, we would expect to incorrectly reject the null of the unit root process 5% of the time (if using the standard 0.05 significance level).

We repeat this experiment 100 times, each time generating a new realization, applying the Dickey-Fuller test, and recording whether the null hypothesis of a unit root process is rejected at the 0.05 significance level. The null hypothesis of a unit root process is rejected 6% of the time. This is approximately correct. Next, we generate a realization of the following AR(1), $\alpha_1 = 0.95$, data-generating process:

$$y_t = 0.95y_{t-1} + \varepsilon_t.$$

This is done using the same generated values for ε_t, as used in the previous experiment. We define $y_1 = \varepsilon_1, y_2 = 0.95y_1 + \varepsilon_2, y_3 = 0.95y_2 + \varepsilon_3$, and so on. The resulting series is a single realization, $T = 300$, of an AR(1) process (not unit root), and ideally, the Dickey-Fuller test would indicate that we can reject the null hypothesis of a unit root process. We repeat this experiment 100 times, each time generating a new realization, applying the Dickey-Fuller test, and recording if the null hypothesis of a unit root process is rejected at the 0.05 significance level.

The result is that the null hypothesis of a unit root process is correctly rejected only 58% of the time. The remaining 42% of the time, the test is unable to reject the null hypothesis of a unit root process even though the process is not unit root. The Dickey-Fuller test appears to have little power to reject the null hypothesis of a unit root process under these conditions.

If we again repeat this 100 times but only use the last 100 time points, we find that the null hypothesis of a unit root process is correctly rejected only 12% of the time. The Dickey-Fuller test has even less power under these conditions. Things improve marginally if we again repeat this 100 times but this time generate data from an AR(1) data-generating process with $\alpha_1 = 0.90$ and T = 100. This time, the null hypothesis of a unit root process is correctly rejected 31% of the time. This is better, but clearly the Dickey-Fuller test has very little power to reject the null hypothesis of a unit root process when

1. the data-generating process is AR(1) with α_1 close to 1 and
2. the number of time points in the data are relatively small.

This must be kept in mind when interpreting the results of a Dickey-Fuller test.

One possible solution is to find or create a data set with a large T that does not necessarily include all the variables in your model but does contain the variables that may or may not be I(1). This can then be used for the purpose of testing if the variables of interest are I(1). With an inferential leap, this can be used to inform the model estimated using the shorter data set. Alternatively, we may have strong theoretical reasons to believe that a data-generating process is AR(1), as we often do with processes like public opinion—in which case, it may be wise to ignore the results of the Dickey-Fuller test unless we have a relatively large sample size.

It should also be noted that some time series are naturally bounded. For example, the proportion of U.S. survey respondents who identify with the Democratic Party is bounded by 0 and 1. Neither the AR(1) process, with or without a trend, nor the I(1), with or without drift or trend, is completely appropriate for bounded series, but the interpretations of the dynamics of each are quite different, and one might be more theoretically appropriate than the other. For example, the I(1) process assumes no equilibrium. This is not likely the case with partisanship over the short term (MacKuen, Erikson, & Stimson, 1989), assuming we control for structural breaks. One empirical approach to this issue is to transform the values of the time series using a logit transformation. If P_t is a time series of proportions, the logit-transformed series is $\ln[P_t / (1 - P_t)]$. Such a series is not bounded.

We could also apply a unit root test that has stationarity as the null hypothesis. One example is the Kwiatkowski-Phillips-Schmidt-Shin (KPSS) test for stationarity (Kwiatkowski, Phillips, Schmidt, & Shin, 1992). The KPSS test can be conducted under the null of either trend stationarity or level (no trend) stationarity. However, the KPSS test also has little power to reject the null, which often leaves us unable to reject either the unit root or the stationarity hypothesis.

A more recent advancement is the Dickey-Fuller generalized least squares test (GLS). The Dickey-Fuller GLS test is the same as the Dickey-Fuller test except that the data are first transformed using a GLS regression. This is done just to give the test greater power (but only a little). We will come back to the Dickey-Fuller GLS test later.

It is also useful to test the null hypothesis that our data come from a random walk with drift process:

$$y_t = y_{t-1} + \alpha_0 + \varepsilon_t. \tag{6.2.4}$$

If we do have such a data-generating process, first differencing produces a stationary, nontrending process: $\Delta y_t = \alpha_0 + \varepsilon_t$. While a random walk with drift process will appear to trend, it would be inappropriate to partial out a trend, like we would with a trend-stationary process. First differencing is the appropriate procedure. Therefore, it is necessary to distinguish between a trend-stationary and a random walk with drift process. We can begin by testing for a unit root process with drift using a variation on the Dickey-Fuller test. The test proceeds by estimating

$$\Delta y_t = \theta y_{t-1} + \alpha_0 + \varepsilon_t. \tag{6.2.5}$$

This is the same regression as Equation 6.2.3, but because the null is a unit root with drift, we test that $\theta = 0$, $\alpha_0 \neq 0$ (Hamilton, 1994). The process under the null hypothesis is a random walk with drift. The alternative is again a stationary process. This is called an augmented Dickey-Fuller test.[3] If we would like to test the null hypothesis of a unit root process (with or without drift) against the alternative of a trend-stationary process, we estimate

$$\Delta y_t = \theta y_{t-1} + \alpha_0 + \beta_1 t + \varepsilon_t$$

and test the null hypothesis: $\theta = 0$ and $\beta_1 = 0$. The null hypothesis is a unit root process (with or without drift).

For our next example, let's look at quarterly data on trust. Trust is operationalized as the percentage of respondents from quarterly surveys indicating that they felt "you can generally trust people." Examining the trust variable in Figure 6.3, we can see that there does appear to be a trend or a drift.

We can use the augmented Dickey-Fuller test to test the null hypothesis that the trust variable has a (a) unit root or (b) unit root with drift data generating process. Beginning with the first, the test statistic is −3.79, with

---

[3] For a more detailed procedure to follow when testing the null hypothesis of a unit root process, see Enders (2004).

174

Figure 6.3 Percentage of Respondents Indicating That You Can
Generally Trust People

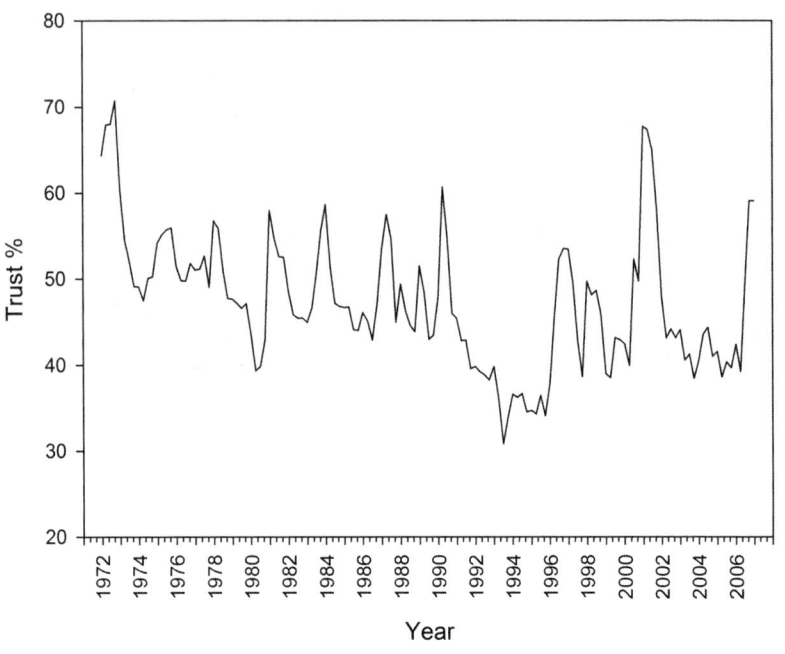

a corresponding P value of 0.003. It appears that we can reject the null of a unit root process.

For the augmented Dickey-Fuller test of a unit root process with drift, the test statistic does not change; rather, it is the critical values that change. Now the corresponding P value is <0.001. It appears that we can also reject the null hypothesis of a unit root process with drift. In both cases, the alternative was a stationary AR(1) process. For the augmented Dickey-Fuller test of a unit root process (with or without drift), against the alternative of a trend-stationary process, the P value is now 0.020. Allowing for the alternative of a trend-stationary process, we can once again reject the null hypothesis of a unit root process.

It is important to note that the Dickey-Fuller test (with stationarity as the alternative) is unlikely to reject the null hypothesis of a unit root if the series is trend stationary. Therefore, if a data series is clearly trending or drifting, we will want to test against the alternative of a trend-stationary data-generating process by using the *augmented* Dickey-Fuller test, or we might mistake a trend-stationary process for one with a unit root.

It is also important to note that unit root tests are less likely to reject the null hypothesis of a unit root, when they should, if the series contains a structural break (Campos, Ericsson, & Hendry, 1996). Therefore, if the series of interest clearly contains an equilibrium shift, this should be partialled out before unit root testing is conducted.

There is a possible alternative to the unit root/stationary dichotomy. It is possible that the process is fractionally integrated. This concept has been gaining popularity since it was originally introduced (Granger & Joyeux, 1980). A fractionally integrated time series process is one that is neither I(0) stationary nor I(d) integrated for any integer value of d. It is an integrated process of a noninteger value of d. This means that it can be transformed into a stationary series through a form of differencing called fractional differencing, just as an integrated series can be transformed into a stationary series through differencing. This is an advanced topic, and the interested reader is directed to the following advanced readings on fractional integration: Granger (1980), Granger and Joyeux (1980), Box-Steffensmeier and Smith (1996, 1998), Lebo and Moore (2003), and Pickup (2009).

Note that we have been using an AR(1) process (with or without a trend) as the alternative when testing for a unit root (with or without drift). The augmented Dickey-Fuller test and KPSS test can also be applied with other stationary processes as the alternative. We can add p lags of Δy_t to allow for more dynamics in the process. The lags are intended to account for additional dynamics. If we use too few, the test will not be right. Returning to our trust example, we apply the augmented Dickey-Fuller test, with a trend-stationary process as the alternative hypothesis, and add three lags of Δy_t. Table 6.1 provides the results from the regression used to produce the augmented Dickey-Fuller test statistic.

The statistical significance of the first lag (of the first difference) suggests that it is a good idea to include it and that the alternative to a unit root (with or without drift) data-generating process is an AR(2) process. The Dickey-Fuller test statistic is −4.46, with a corresponding P value of 0.002. We can still reject the null hypothesis of a unit root process against the alternative of a trend-stationary process.

The only difficulty with including additional lags of the first difference is that it reduces the number of time points available; remember that the Dickey-Fuller (and KPSS) test is already problematic without a long time series of data. A solution to this problem is the Phillips-Perron test for a unit root (Phillips & Perron, 1988). This is the same as the Dickey-Fuller test but uses Newey-West heteroskedasticity—and autocorrelation—consistence standard errors, instead of including lags of the first difference dependent variable.

As with the Dickey-Fuller test, we can also specify the alternative hypothesis of a trend-stationary process. Specifying standard errors that are

Table 6.1 Regression Table for Dickey-Fuller Test

Trust	Coefficient	Standard Error	t Statistic	P Value
L1. Trust	−0.29	0.065	−4.46	0.00
LD. Trust	0.23	0.086	2.65	0.009
L2D. Trust	0.095	0.087	1.09	0.276
L3D. Trust	−0.081	0.088	−0.92	0.361
Trend	−0.011	0.011	−1.07	0.286
Constant	14.34	3.52	4.07	0.00

NOTE: L1 = first lag, L2 = second lag, L3 = third lag, D = difference.

corrected for up to four lags of serial correlation, the Phillips-Perron tau test statistic is −3.89, with a corresponding P value of 0.013. Based on these results, we can again reject the null hypothesis of a unit root process against the alternative of a trend-stationary process.

The fact is that we often do not have enough data to reject a unit root process, in which case we may need to rely on evidence from other similar but longer time series to decide. If we truly do believe that we have a unit root data-generating process, then we will want to take account of this. We have seen that first differencing our data is one approach. We turn to models that use differenced data next, but first we consider higher-order unit root processes.

It is possible to test for higher-order unit root processes with the tests we have discussed, such as the Dickey-Fuller test. For example, y_t is an I(2) process if it is stationary after being differenced twice. If we apply the Dickey-Fuller test to the first difference of y_t, we are testing the null hypothesis of a second order unit root process I(2) against the alternatives of a first order unit root process I(1) or a stationary process—see Enders (204, pp. 194-195) for details. There are more powerful (and complex) ways of testing for higher-order unit root processes (e.g., Dickey & Pantula, 2002), but it is rare within the social sciences to go beyond testing for an I(1) process.

6.3 Differenced Data and Autoregressive Integrated Moving Average (ARIMA) Models

This brings us next to two new data models. If the dependent variable has an I(1) data-generating process, we can use what is called a differenced data model. Starting with the autoregressive distributed lag model, ADL(1,1),

$$y_t = \alpha_0 + \alpha_1 y_{t-1} + \beta_1 x_t \mp \beta_2 x_{t-1} + \varepsilon_t. \tag{6.3.1}$$

If we believe that y_t is I(1), $\alpha_1 = 1$, this is an inappropriate model, but Equation 6.3.1 with $\alpha_1 = 1$ can be transformed into a differenced data model. Subtract y_{t-1} from each side, and assume that $\beta_2 = -\beta_1$; then,

$$y_t - y_{t-1} = \alpha_0 + y_{t-1} - y_{t-1} + \beta_1 \left(x_t - x_{t-1} \right) + \varepsilon_t,$$

$$\Delta y_t = \alpha_0 + \beta_1 \Delta x_t + \varepsilon_t. \tag{6.3.2}$$

This is a differenced data model (Hendry, 2003). If y_t is I(1), Δy_t is I(0); and this model is appropriate for estimation by OLS or maximum likelihood estimation. Note that x_t is also first differenced.[4] What does it mean to assume that $\beta_2 = -\beta_1$ in this context? It assumes that a permanent one-unit change in x_t at t has an effect at t, and that is the sum total of the effect—like a static model. This type of approach has been suggested as appropriate for consumption models (Campbell & Mankiw, 1991).

After estimating a differenced data model, we might find that the residuals are serially correlated. This motivates our next model: An ARIMA model is simply an ARMA model with the dependent and independent variables differenced one or more times.

The purpose of an ARIMA model is simply to difference transform a variable with an I(d) data-generating process into one with an I(0) data-generating process and then apply the usual approach to building an ARMA model—including autoregressive and moving average components. If it does not contain any autoregressive or moving average components, then it is a differenced data model. In an ARIMA model, we may also difference the data to account for seasonality. Sometimes such seasonal ARIMA models are denoted by the acronym SARIMA.

Let us say that we wish to run a difference data model on the trust variable from our previous example, with civic engagement as an independent variable.[5] We specify an ARIMA with zero autoregressive components,

---

[4] Modelling Δy_t as a function of x_t would be to model a process in which any nonchanging, nonzero value for x_t would cause y_t to steadily and perpetually increase or decrease.

[5] Data are from Keele (2007). The civic engagement variable is based on four indicators: (1) participation in community organizations, (2) participation in politics and public affairs, (3) volunteering, and (4) informal socializing. See Keele (2005, 2007) for further details.

178

zero moving average components, and once differencing of the variables as follows:

$$\Delta\text{trust}_t = \alpha_0 + \beta_1 \Delta\text{engage}_t + \varepsilon_t.$$

In other words, we have a differenced data model. Before running this model, we examine the autocorrelation function (ACF) and the partial

Figure 6.4   ACF and PACF for Differenced Trust

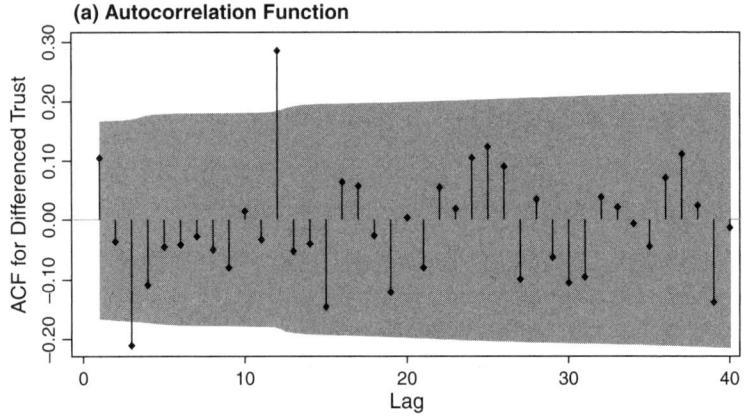

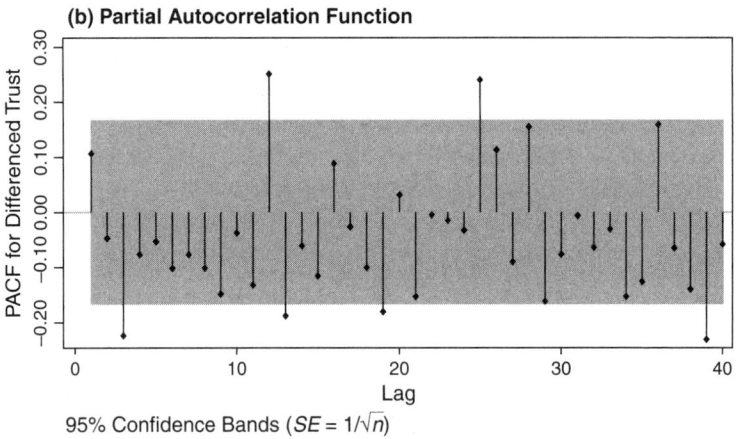

NOTE: ACF = autocorrelation function, PACF = partial autocorrelation function, MA = moving average, *SE* = standard error.

Table 6.2 ARIMA ($d = 1$, $q = 3$) Model of Trust

D1. Trust	Coefficient	Standard Error	z Statistic	P Value
D1. Engage	0.70	0.29	2.43	0.015
Constant	−0.087	0.31	−0.28	0.780
L3. MA	−0.29	0.11	−2.56	0.010

NOTE: Log likelihood = −393.439, $T = 136$; D1 = first difference, L3 = third lag, ARIMA = autoregressive integrated moving average model, MA = moving average, P = probability, T = number of time points.

autocorrelation function (PACF) for $\Delta trust_t$. These are presented in Figure 6.4.

Now let us say we estimate an ARIMA, based on Figure 6.4, with the independent and dependent variables differenced once, zero autoregressive components, and an MA($q = 3$) component. The results are presented in Table 6.2.

For the residuals, we estimate the ACF and PACF (Figure 6.5) and calculate the Q statistic.

The portmanteau (Q) statistic is 47.81 and is chi-squared distributed with 40 degrees of freedom. The corresponding P value is 0.185. Therefore, we cannot reject the null hypothesis of a white noise process for the residuals from this model.

This model assumes that there is no long-run effect of civic engagement on trust, beyond the short-run effect, in the data-generating process (or at very least, we do not model it). If this is not true, the data model is misspecified, and the inference regarding the significance of the parameters will be incorrect or incomplete. We will also not properly understand the relationship between trust and civic engagement. These issues will motivate the next model: the ECM.

6.4 Cointegration and the Error Correction Model (ECM)

Before discussing ECMs, we need to understand the concept of cointegration. Say for two I(1) data-generating processes, y_t and x_t, there is a β such that $s_t = y_t - \beta x_t$ is a stationary process (i.e., an I(0) process). If so, we say that y_t and x_t are cointegrated and call β the cointegration parameter (Engle & Granger, 1987). We denote this relationship C(1,1), meaning that y_t and x_t are I(1) and cointegrated to produce a series that is one order of integration lower—that is, I(0). If we know β, testing for cointegration is straightforward.

180

Figure 6.5   ACF and PACF for Residuals From ARIMA ($d = 1$, $q = 3$)
Model of Trust

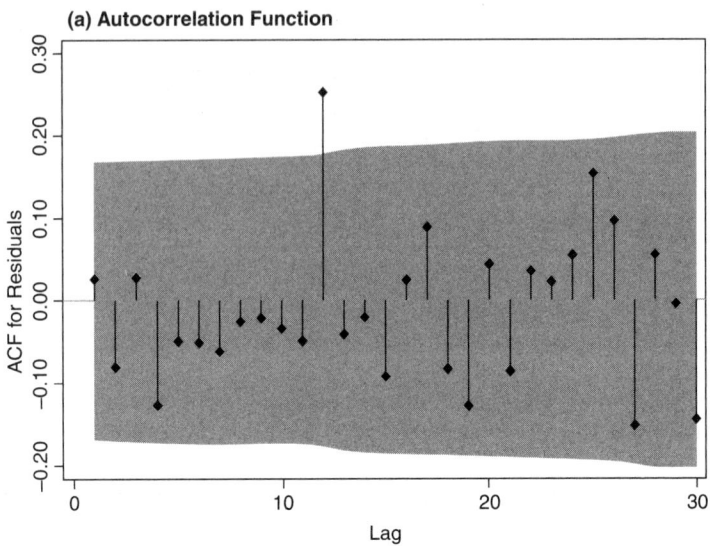

(a) Autocorrelation Function

Bartlett's Formula for MA(q) 95% Confidence Bands

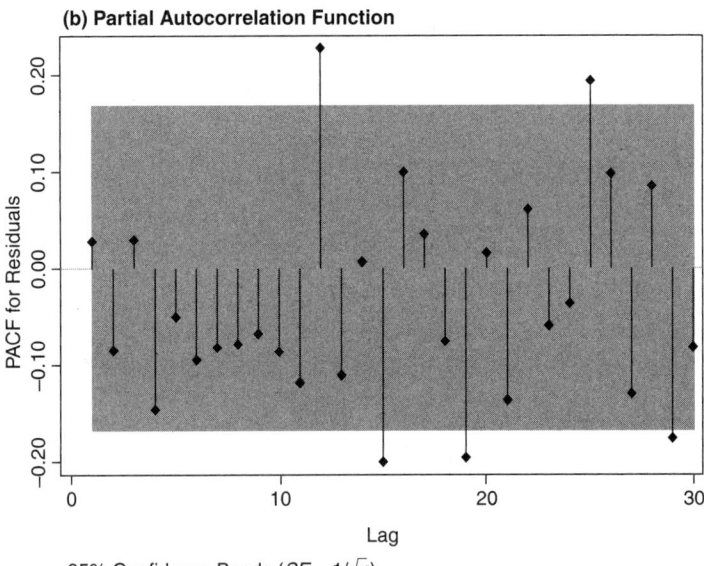

(b) Partial Autocorrelation Function

95% Confidence Bands ($SE = 1/\sqrt{n}$)

NOTE: ACF = autocorrelation function, PACF = partial autocorrelation function, ARIMA = autoregressive integrated moving average model, MA = moving average, SE = standard error.

1. First confirm that y_t and x_t are both I(1), using the Dickey-Fuller or equivalent test.

2. Second, define $s_t = y_t - \beta x_t$. We could also add a constant, trend, periodicity, or structural break to this if we believed that it was part of the cointegrating process. For example, $s_t = y_t - \alpha - \beta x_t$.

3. Third, run the Dickey-Fuller or equivalent test on s_t and if we reject a unit root, then y_t and x_t are cointegrated.

Of course, β is generally unknown, as is any constant, trend, periodicity, or structural break, so we first have to estimate these, which adds a complication. We can estimate $y_t = \beta x_t + \mu_t$, and calculate

$$\hat{s}_t = \hat{\mu}_t = y_t - \hat{\beta} x_t. \tag{6.4.1}$$

And we can apply the Dickey-Fuller test to $\hat{s}_t$, in order to determine if it is I(0), but this requires different critical values from those for the usual Dickey-Fuller test. A higher threshold is needed to take into account that the β is estimated and not known ahead of time, and therefore, the Dickey-Fuller test is applied to estimates of s_t. The appropriate critical values can be looked up (MacKinnon, 2010).

If we believe that it is part of the cointegrating relationship, we can include a constant or trend to the initial regression that estimates β, and use different critical values again when testing whether $\hat{s}_t$ is I(0). The common practice is to include a constant: $\hat{s}_t = y_t - \hat{\alpha} - \hat{\beta} x_t$. We shall revisit the issue of including constants or trends in cointegrating relationships shortly.

If the two I(1) processes are cointegrated, β tells us the long-run relationship between y_t and x_t. We will return to this later. If the two I(1) processes are not cointegrated, then $y_t = \hat{\alpha} - \hat{\beta} x_t + \hat{\mu}_t$ is a spurious regression, the function $s_t = y_t - \alpha - \beta x_t$ has no meaning, and $\hat{\beta}$ tells us nothing.

Let us introduce a new example: the relative defense expenditures of the United States and U.S.S.R. These are yearly defense expenditures in U.S. dollars from 1967 to 1988 (Ostrom, 1990; Ostrom & Marra, 1986). They are plotted in Figure 6.6.

As expenditures are drifting/trending upward, we begin by testing the U.S. defense expenditures against the null hypothesis of a unit root process with or without drift, including a trend in the alternative hypothesis. The test statistic for the augmented Dickey-Fuller test is -1.097, with a corresponding P value of 0.93. It would seem that we cannot reject the null hypothesis of a unit root process against the alternative of a trend-stationary process, but let us also test the null hypothesis of a unit root process with drift against the alternative of a stationary process without the trend.

182

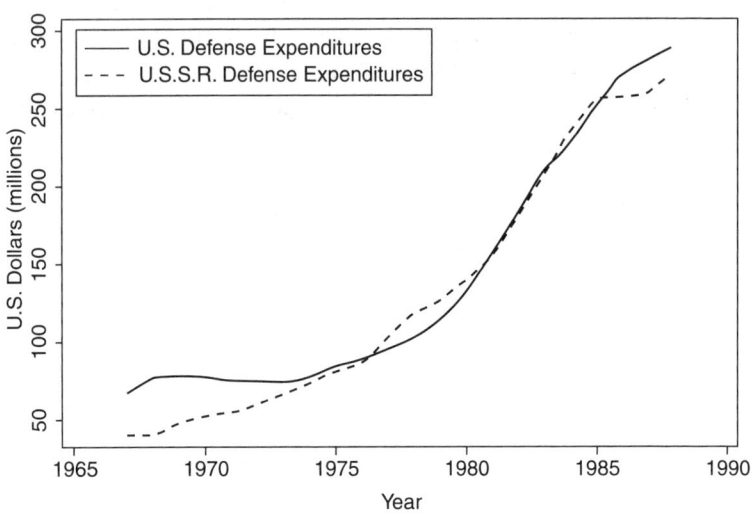

Year

The test statistic for the augmented Dickey-Fuller test is 2.87, with a corresponding P value of 0.995. It appears that we cannot reject the null hypothesis of a unit root process with drift. We do the same thing for U.S.S.R. defense expenditures.

Testing the null hypothesis of a unit root process against the alternative of a trend-stationary process, the test statistic for the augmented Dickey-Fuller test is -1.85, with a corresponding P value of 0.68. Testing the null hypothesis of a unit root process with drift, the test statistic for the augmented Dickey-Fuller test is 1.82, with a corresponding P value of 0.958. Again, it would seem that we cannot reject the null hypothesis of a unit root process with drift and we cannot reject the null hypothesis of a unit root process against the alternative of a trend-stationary process.

On the basis that we believe that both U.S. and U.S.S.R. defense expenditures are unit root processes (with drift), we regress U.S. expenditures on U.S.S.R. expenditures and request the residuals. But first we reexamine the data in Figure 6.6. As noted, both series seem to contain a drift or trend. When testing each series against the hypothesis of a unit root process, it was necessary to control for this drift/trend. This does not necessarily mean that there is a trend within the cointegrating relationship. In fact, a strict definition of cointegration requires the cointegrating relationship to produce a stationary series, ruling out the possibility of a trend. However, this can be relaxed to allow the cointegrating relationship to produce a trend-stationary series.

For our current example, we include a constant in the cointegrating relationship (as is the convention) and choose to be agnostic about the trend by also including it.

$$\mathrm{US}_t = \lambda + \kappa_1 \mathrm{USSR}_t + \alpha_1 \mathrm{Year}_t + \mu_t. \tag{6.4.2}$$

The results from the OLS estimation of this model are presented in Table 6.3. This gives us the potential cointegrating relationship. From this cointegrating relationship, we predict residuals and test for cointegration.

$$\hat{s}_t = \hat{\mu}_t = \mathrm{US}_t - \hat{\lambda} - \hat{\kappa}_1 \mathrm{USSR}_t - \hat{\alpha}_1 \mathrm{Year}_t \tag{6.4.3}$$

The Dickey-Fuller test statistic for the residuals is -2.35. The critical value for cointegration, including a deterministic trend, at the 0.05 significance level is -4.24 (MacKinnon, 2010), as the test statistic is not lower than the critical value (left-tailed test). It appears that we cannot reject the null of a unit root process in the residuals, and therefore, we cannot reject the hypothesis that the variables are not cointegrated. In other words, this test does not provide evidence of cointegration. However, we shall find later in this chapter that this test is no longer the preferred test for cointegration and the preferred test does find evidence of cointegration.[6]

Table 6.3 Cointegrating Relationship Between U.S. and U.S.S.R. Defense Expenditures

U.S. Defense Expenditures	Coefficient	Standard Error	t Statistic	P Value
U.S.S.R. Defense Expenditures	1.42	0.11	13.33	0.000
Year	−6.22	1.35	−4.60	<0.001
Constant	12,261.35	2,662.69	4.60	<0.001

NOTE: $R^2 = 0.99$, $T = 22$; $T =$ number of time points.

---

[6] It has been noted that a series may be fractionally integrated. It follows that two or more series may be fractionally cointegrated. Again, this is an advanced topic. See Davidson (2002) and Cheung and Lai (1993) for a review of the issues encountered when testing for fractional cointegration.

Error Correction Models

Cointegration allows us to expand the type of dynamic models that can be estimated for I(1) data beyond using first differences. Since $\hat{s}_t$ is stationary, it can be included in time series models. Such models include ECMs. They are also known as *equilibrium correction* models (Hendry, 2003). Unlike differenced data models, ECMs allow us to examine both long-run and short-run relationships.

A standard ECM is just an ADL(1,1) model that has been transformed as follows. Starting with the ADL(1,1) data-generating process,

$$y_t = \alpha_0 + \alpha_1 y_{t-1} + \beta_1 x_t + \beta_2 x_{t-1} + \varepsilon_t. \tag{6.4.4}$$

Subtracting y_{t-1} from each side of Equation 6.4.4, we get

$$y_t - y_{t-1} = \alpha_0 + (\alpha_1 - 1)y_{t-1} + \beta_1 x_t + \beta_2 x_{t-1} + \varepsilon_t,$$

$$\Delta y_t = \alpha_0 + (\alpha_1 - 1)y_{t-1} + \beta_1 x_t + \beta_2 x_{t-1} + \varepsilon_t. \tag{6.4.5}$$

Adding and subtracting $\beta_1 x_{t-1}$, on the right-hand side of Equation 6.4.5, we get

$$\Delta y_t = \alpha_0 + (\alpha_1 - 1)y_{t-1} + \beta_1 x - \beta_1 x_{t-1} + \beta_1 x_{t-1} + \beta_2 x_{t-1} + \varepsilon_t.$$

Note that

$$\beta_1 x_t - \beta_1 x_{t-1} = \beta_1 \Delta x_t \text{ and } \beta_1 x_{t-1} + \beta_2 x_{t-1} = (\beta_1 + \beta_2)x_{t-1},$$

$$\Delta y_t = \alpha_0 + (\alpha_1 - 1)y_{t-1} + (\beta_1 + \beta_2)x_{t-1} + \beta_1 \Delta x_t + \varepsilon_t. \tag{6.4.6}$$

Next, use

$$(\beta_1 + \beta_2) = \frac{(\alpha_1 - 1)(\beta_1 + \beta_2)}{(\alpha_1 - 1)} = -\frac{(\alpha_1 - 1)(\beta_1 + \beta_2)}{(1 - \alpha_1)}.$$

And insert into Equation 6.4.6

$$\Delta y_t = \alpha_0 + (\alpha_1 - 1)y_{t-1} - \frac{(\alpha_1 - 1)(\beta_1 + \beta_2)}{(1 - \alpha_1)}x_{t-1} + \beta_1 \Delta x_t + \varepsilon_t. \tag{6.4.7}$$

Collect the terms multiplied by $(\alpha_1 - 1)$:

$$\Delta y_t = \alpha_0 + (\alpha_1 - 1)\left[y_{t-1} - \frac{(\beta_1 + \beta_2)}{(1 - \alpha_1)}x_{t-1} \right] + \beta_1 \Delta x_t + \varepsilon_t. \tag{6.4.8}$$

This transformation motivates the standard ECM:

$$\Delta y_t = \alpha_0 + \gamma \left(y_{t-1} - \kappa_1 x_{t-1} \right) + \kappa_0 \Delta x_t + \varepsilon_t, \qquad (6.4.9)$$

where $\gamma \equiv (\alpha_1 - 1)$, $\kappa_0 \equiv \beta_1$, and $\kappa_1 \equiv (\beta_1 + \beta_2)/(1 - \alpha_1)$.

As we will discuss later in the chapter, ECMs can be applied to stationary data—in which case, κ_0 is equivalent to the ADL short-run impact of a unit change in x_t, and κ_1 is equivalent to the ADL long-run impact; and this is exactly how they are interpreted. This standard ECM models an ADL(1,1) process where the short- and long-run effects are explicitly modelled as parameters. However, if y_t and x_t are C(1,1), the long-run effect term is the cointegrating relationship between them:

$$y_{t-1} - \kappa_1 x_{t-1} = s_{t-1},$$

which is a different type of long-run relationship. In this long-run relationship, two I(1) variables track each other in such a way that the (proportional) difference is a stationary process. If y_t and x_t are I(1), the remaining terms in the ECM (Δy_{t-1} and Δx_{t-1}) are stationary processes. Therefore, all constituent processes in the ECM are stationary (as long as y_t and x_t are cointegrated). The constant in Equation 6.4.9 can be moved into the cointegrating relationship, and the ECM can be rewritten as follows:

$$\Delta y_t = \gamma \left(y_{t-1} - \lambda - \kappa_1 x_{t-1} \right) + \kappa_0 \Delta x_t + \varepsilon_t, \qquad (6.4.10)$$

where $\lambda \equiv -(\alpha_0/\gamma) = \alpha_0/(1 - \alpha_1)$, which is in the estimate of the ADL(1,1) equilibrium if the data are stationary.

A nonzero constant outside the cointegrating relationship, as in Equation 6.4.9, can imply a linear trend in the levels of y_t, just as it does in the random walk with drift process. It may, however, imply that a constant should be included in the cointegrating relationship (Equation 6.4.10). In fact, the data-generating process may contain both. This possibility regarding the data-generating process is true regardless of whether or not the constant is included inside (Equation 6.4.10) or outside (Equation 6.4.9), the cointegrating relationship in the data model. It makes theoretical sense to include both if you believe that the data-generating process includes both. However, such a model is not identified, and some identifying strategy is required. Such a strategy is left to a more advanced text (Johansen, 1988; Maddala & Kim, 1998,) but a general guide is to include the constant (drift term) outside the cointegrating relationship if y_t exhibits a trend and to include it within the cointegrating relationship otherwise.

Just as a constant can be included inside or outside the cointegrating relationship, a trend term can be included inside or outside the cointegrating relationship. However, they imply different data-generating processes. A trend within the cointegrating relationship is appropriate if y_t and x_t are cointegrated, controlling for the linear trend. A trend outside the cointegrating relationship is appropriate if Δy_t contains a trend that would exhibit itself as a quadratic trend in y_t.

If the data are cointegrated, the ECM can be estimated using the Engle-Granger two-step procedure (Engle & Granger, 1987). For example, if we were applying this to the economic popularity model we have seen a number of times already, including for the moment only gross domestic product (GDP) as a covariate, the ECM would be as follows:

$$\Delta \text{Pop}_t = \gamma \left(\text{Pop}_{t-1} - \lambda - \kappa_1 \text{GDP}_{t-1} \right) + \kappa_0 \Delta \text{GDP}_t + \varepsilon_t.$$

The Engle-Granger two-step procedure would proceed as follows:

In Step 1, estimate by OLS $\quad \text{Pop}_t = \lambda + \kappa_1 \text{GDP}_t + \mu_t.$

In Step 2, calculate $\quad \hat{s}_t = \hat{\mu}_t = \text{Pop}_t - \hat{\lambda} - \hat{\kappa}_1 \text{GDP}_t.$

And by OLS, estimate $\quad \Delta \text{Pop}_t = \gamma \hat{s}_{t-1} + \kappa_0 \Delta \text{GDP}_t + \varepsilon_t.$

Unfortunately, we cannot use the standard errors from Step 1 for hypothesis testing (e.g., Is the coefficient statistically different from 0?) for κ_1 because the variables are unit root. This is not a problem when estimating the coefficient of cointegration, but it is when testing the statistical significance of κ_1. The solution is to use the regression results from Step 1 to estimate $\hat{s}_t$, but to use a separate procedure to test the significance of κ_1.

One available technique is the leads and lags estimate. This technique involves reestimating the regression from Step 1 with the addition of the first difference of Δx_t (in this case ΔGDP_t) and both leads and lags of Δx_t: $\Delta x_{t+1}, \Delta x_{t-1}, \Delta x_{t+2}, \Delta x_{t-2}, \dots$. Leaving the rationale for an advanced text (Stock & Watson, 1993), the inclusion of enough leads and lags produces unbiased estimates of the standard errors for $\hat{\kappa}_1$. The number of leads and lags to include is going to be restricted by the data. If we have only a few data points, each additional lead and lag increases our standard errors; however, a greater number of leads and lags means greater confidence in our results. A single lead–lag estimate would proceed as follows. (1) Estimate

$$\text{Pop}_t = \lambda + \kappa_1 \text{GDP}_t + \psi_0 \Delta \text{GDP}_t + \psi_1 \Delta \text{GDP}_{t+1} + \psi_2 \Delta \text{GDP}_{t-1} + \mu_t.$$

The ψ_i values are just parameters to be estimated. (2) Use the new estimate of κ_1 and the new estimate of its standard errors to test the null hypothesis $\kappa_1 = 0$.

We return to our U.S. and U.S.S.R defense expenditures example.[7] As noted previously, we will learn later in this chapter that there is evidence that the U.S. and U.S.S.R. defense expenditures are cointegrated $C(1,1)$. Anticipating that evidence, we use the Engle-Granger two-step procedure to estimate an ECM using the U.S. and U.S.S.R. data. The results presented in Table 6.3 represent Step 1, giving us

$$US_t = \hat{\lambda} + \hat{\kappa}_1 USSR_t + \hat{\alpha}_1 \, Year_t + \hat{\mu}_t,$$

from which we can estimate the residuals:

$\hat{s}_{t-1} = US_t - \hat{\lambda} - \hat{\kappa}_1 USSR_t - \hat{\alpha}_t \, Year_t$. In Step 2, we estimate the ECM:

$$\Delta US_t = \gamma \hat{s}_{t-1} + \kappa_0 \Delta USSR_t + \varepsilon_t.$$

Note that we have not included a constant in this step, as we have included it in the cointegrating relationship.

The results in Table 6.4 give us the immediate effect of U.S.S.R. expenditures on U.S. expenditures: $\hat{\kappa}_0 = 0.86$. This effect is significant at the 95% confidence level. We can also test the residuals from the ECM against the null hypothesis of a white noise process. The portmanteau Q statistic is 15.10 and is chi-squared distributed with 8 degrees of freedom. The corresponding P value is 0.057. Therefore, we cannot reject the null hypothesis of a white noise process for the residuals from this model at the 0.05 significance level.

Note that the estimated coefficient of cointegration for U.S.S.R. and U.S. expenditures is $\hat{\kappa}_1 = 1.42$ (Table 6.3), but we cannot use the cointegrating regression to test the significance of this coefficient. To do this, we estimate the lead–lag estimator of the cointegrating relationship, including one lead and one lag.

The result (Table 6.5) is that the coefficient of cointegration for U.S.S.R. and U.S. expenditures does appear to be significant at the 0.05 significance level. Note, however, that the estimated coefficient from the lead–lag estimator is different from that from the original OLS regression: $\hat{\kappa}_1 = 1.61$. This is a problem with the Engle-Granger two-step procedure. It has been demonstrated that over repeated samples, there will be no difference

---

[7] For an extensive discussion of the value of error correction models within the discipline of political science, see Durr (1992) and the other articles in the same issue of *Political Analysis*.

between the coefficient from the lead–lag estimator and that from the OLS estimator (Engle & Granger, 1987), but of course, this is not true in any given sample.

The results from Table 6.5 show us that U.S.S.R. defense expenditures have a statistically significant effect within the cointegrating relationship. For the U.S.S.R. defense expenditures to have a statistically significant effect on U.S. defense expenditures, it is also necessary that $\hat{s}_{t-1}$ has a statistically significant coefficient within the main ECM. The results presented in Table 6.4 suggest that this is the case at the 0.05 significance level. The cointegrating relationship is a type of long-run equilibrium between x_t and y_t. Therefore, the effect of the cointegrating relationship on US expenditures indicates that US expenditures respond to US and USSR expenditures being out of their equilibrium, so as to bring the two back into equilibrium—see Enders (2010) for a further discussion of interpreting ECMs.

Table 6.4 U.S. and U.S.S.R. Defense Expenditure Error Correction Model

D1. U.S. Defense Expenditures	Coefficient	Standard Error	t Statistic	P Value
D1. U.S.S.R. defense expenditures	0.86	0.096	8.89	<0.001
L1. Cointegrating relationship	−0.48	0.15	−3.22	0.005

NOTE: $R^2 = 0.59$, $T = 21$; T = number of time points, D1 = first difference, L1 = first lag.

Table 6.5 U.S. and U.S.S.R. Defense Expenditure Error Correction Model

U.S. Defense Expenditures	Coefficient	Standard Error	t Statistic	P Value
Year	−8.36	0.95	−8.78	<0.001
U.S.S.R. defense expenditures	1.61	0.068	23.62	<0.001
D	−0.96	0.22	−4.44	<0.001
FD	0.76	0.20	3.86	0.002
LD	−0.50	0.21	−2.35	0.035
Constant	16,469.93	1,874.57	8.79	<0.001

NOTE: $R^2 = 0.996$, $T = 19$; P = probability, T = number of time points, D = first difference, FD = first lead of the first difference, LD = first lag of the first difference.

The simple ECM examined so far can be extended in a number of ways. First, additional variables can be added to the right-hand side. For example, if $y_{t-1} - \lambda - \kappa_{1,1} x_{t-1} - \kappa_{1,2} z_{t-1}$ represents a cointegrating relationship, our ECM could be as follows:

$$\Delta y_t = \gamma \left(y_{t-1} - \lambda - \kappa_{1,1} x_{t-1} - \kappa_{1,2} z_{t-1} \right) + \kappa_{0,1} \Delta x_t + \kappa_{0,2} \Delta z_t + \varepsilon_t. \quad (6.4.11)$$

We could also add additional lags of Δx_t and/or Δy_t to the ECM. For example,

$$\Delta y_t = \gamma \left(y_{t-1} - \lambda - \kappa_1 x_{t-1} \right) + \kappa_0 \Delta x_t + \kappa_2 \Delta x_{t-1} + \kappa_3 \Delta y_{t-1} + \varepsilon_t. \quad (6.4.12)$$

In doing so, the ECM is no longer isomorphic to the ADL(1,1) model and would not be appropriate for the stationarity ADL(1,1) data-generating process, but it is a perfectly valid model for the corresponding data-generating process.

We are now in a position to discuss more recent developments in time series analysis involving cointegration. Such developments have occurred in the context of vector error correction models (VECMs). This topic is beyond the scope of this book, but a brief introduction will allow us to understand modern approaches to testing cointegration. Start with the following two error correction processes:

$$\Delta y_t = \gamma_1 \left(y_{t-1} - \lambda_1 - \kappa_{1,1} x_{t-1} \right) + \varepsilon_{1t},$$

$$\Delta x_t = \gamma_2 \left(x_{t-1} - \lambda_2 - \kappa_{2,1} y_{t-1} \right) + \varepsilon_{2t}. \quad (6.4.13)$$

Note that we have had to add an extra index on the cointegrating coefficients to keep them clear. In combination, the above two equations (6.4.13) are known as a VECM.

Just as the standard ECM is isomorphic to the ADL(1,1), the VECM in Equation 6.4.13 is isomorphic to the following model, called a vector autoregression model:

$$y_t = \alpha_{1,0} + \alpha_{1,1} y_{t-1} + \beta_{1,1} x_{t-1} + \omega_{1t},$$

$$x_t = \alpha_{2,0} + \alpha_{2,1} x_{t-1} + \beta_{2,1} y_{t-1} + \omega_{2t}. \quad (6.4.14)$$

The transformation from Equation 6.4.14 back to Equation 6.4.13 is analogous to the transformation from the ADL(1,1) to the standard ECM. Just as in that transformation, it is required that we define $\gamma_1 = \left(\alpha_{1,1} - 1 \right)$ and $\gamma_2 = \left(\alpha_{2,1} - 1 \right)$. Considering just the equation for y_t in Equation 6.4.14, recall that if y_t is unit-root, then $\alpha_{1,1} - 1 = \gamma_1 = 0$. If we reject the null hypothesis, $\gamma_1 = 0$, we conclude that y_t is stationary. If we fail to reject the null hypothesis, we conclude that y_t is unit root.

Returning to the multi-equation case (Equation 6.4.14), we can conduct a test of cointegration by examining the following matrix formed by the parameters in the equation:

$$\begin{bmatrix} \alpha_{1,1} - 1 = \gamma_1 & \beta_{1,1} \\ \beta_{2,2} & \alpha_{2,1} - 1 = \gamma_2 \end{bmatrix}. \qquad (6.4.15)$$

We proceed by estimating the rank of this matrix. The rank is the maximum number of linearly independent rows. Johansen (1995, 1988) demonstrated that if y_t and x_t are cointegrated, the two rows will not be linearly independent and the rank is equal to the number of cointegrating vectors. Also, the maximum number of possible cointegrating equations is the number of rows minus 1. With only two variables, an estimated rank of 2 means that the two rows are independent, which implies that y_t and x_t are both stationary. If the estimated rank is 0 and the matrix is all zeros, we fail to reject both the null hypothesis that $\gamma_1 = 0$ and the null hypothesis that $\gamma_2 = 0$, and by Equation 6.4.13,

$$\Delta y_t = \varepsilon_{1t,}$$

$$\Delta x_t = \varepsilon_{2t} \qquad (6.4.16)$$

We conclude that both y_t are x_t are unit root and are not cointegrated. An estimated rank of 1 indicates that a single cointegrating relationship exists. Note that with only two variables, the maximum number of cointegrating vectors is 1.

Johansen (1995, 1988) provides a maximum likelihood estimate of the parameters in Equation 6.4.14 and two test statistics for determining the rank of the matrix in Equation 6.4.15. We will discuss the maximum eigenvalue statistic (λ_{max}). This statistic can be calculated for each possible rank value. It can first be calculated to test the null hypothesis that the rank is 0 against the alternative that the rank is 1. If we reject the null that the rank is 0 against this alternative, we can calculate the statistic to test the null hypothesis that the rank is 1 against the alternative that the rank is 2. Using this procedure, we now reexamine the potential cointegrating relationship between U.S. and U.S.S.R. defense expenditures. As before, we include a constant and a trend within the cointegrating relationship of the VECM.

In Table 6.6, the row that starts with maximum rank 0 tests the null hypothesis that the rank is 0 against the alternative that the rank is 1. In other words, the null hypothesis is that there is no cointegration. This can be rejected. The row that starts with maximum rank 1 tests the null hypothesis

Table 6.6 Johansen Maximum Eigenvalue Statistics

Maximum Rank	Parms	LL	Eigenvalue	Maximum Statistic	5% Critical Value
0	6	−124.14	—	21.11	18.96
1	10	−113.59	0.65	9.08	12.52
2	12	−109.05	0.36	—	—

NOTE: LL= Loglikelihood.

that the rank is 1 against the null hypothesis that the rank is 2. The null in this case is that there is one cointegrating equation, and the alternative is that both variables are stationary. This null hypothesis cannot be rejected, meaning that we conclude that there is a cointegrating equation. If we had rejected the null in this case, we would have concluded that the two series are stationary. Based on our test results, we can use the cointegrating relationship regression (Equation 6.4.2) to estimate the cointegrating relationship between USSR and US expenditures.

Given that both x_t and y_t can be seen as dependent variables in a cointegrating relationship, we need to ask when it is valid to estimate the single-equation model for Δy_t—as we did in the case of the U.S., U.S.S.R. defense expenditures example. This can be answered by returning to Equation 6.4.13. The requirement for estimating the single-equation ECM for Δy_t is that γ_2 in Equation 6.4.13 is equal to 0. If this is the case, then x_t does not respond to x_t and y_t being out of equilibrium. It is the response of y_t that brings them back into equilibrium. Under these circumstances, we can say that x_t is weakly exogenous for the estimation of the single equation for Δy_t in Equation 6.4.13. Recall that we introduced the concept of weak exogeneity in Chapter 2.

Error Correction Models With Stationary Data

As mentioned previously, ECMs can also be used with data produced by stationary data-generating processes. Consider again government popularity and GDP data with stationary data-generating processes. The standard error correction model is as follows:

$$\Delta \text{Pop}_t = \gamma \left(\text{Pop}_{t-1} - \lambda - \kappa_1 \text{GDP}_{t-1} \right) + \kappa_0 \Delta \text{GDP}_t + \varepsilon_t.$$

Multiplying out the cointegrating relationship, this can be rewritten in the "general form":

$$\Delta \text{Pop}_t = \gamma \text{Pop}_{t-1} - \gamma \lambda - \gamma \kappa_1 \text{GDP}_{t-1} + \kappa_0 \Delta \text{GDP}_t + \varepsilon_t.$$

This can be rewritten as

$$\Delta \text{Pop}_t = \gamma \text{Pop}_{t-1} + \delta_0 + \delta_1 \Delta \text{GDP}_t + \delta_2 \text{GDP}_{t-1} + \varepsilon_t, \qquad (6.4.17)$$

where $\delta_0 = -\gamma\lambda$, $\delta_1 = \kappa_0$, $\delta_2 = -\gamma\kappa_1$. The expression of the standard ECM is easier to interpret because κ_1 is estimated directly, but the general form with stationary data can be estimated using OLS in a single step—this cannot be done with unit root data because of the inclusion of GDP_{t-1}.[8] If Pop_t and GDP_t are stationary, so are ΔPop_t and ΔGDP_t; therefore, we can use OLS to estimate Equation 6.4.17. After estimating γ, δ_0, δ_1, and δ_2, we can calculate the estimated long-run effect $\hat{\kappa}_1 = -(\hat{\delta}_2 / \hat{\gamma})$, and the estimated equilibrium, $\hat{\lambda} = -(\hat{\delta}_0 / \hat{\gamma})$.

Using the German government popularity and economic data that we have used before, we estimate the following ECM (we have three independent variables—GDP, inflation, and unemployment) (Table 6.7):

$$\Delta \text{Pop}_t = \gamma \text{Pop}_{t-1} + \delta_0 + \delta_{1a} \Delta \text{GDP}_t + \delta_{2a} \text{GDP}_{t-1} + \delta_{1b} \Delta \inf_t + \delta_{2b} \inf_{t-1}$$

$$+ \delta_{1c} \Delta \text{unemp}_t + \delta_{2c} \text{unemp}_{t-1} + \varepsilon_t.$$

Note that we also included a trend as we previously determined that the German government popularity variable is a trend-stationary data-generating process. The trend does not need to be entered as its first difference or lag because the first difference of a trend is simply a constant and the lag of a trend is just a trend.

The estimated coefficient on each of the first-differenced economic variables is the corresponding $\hat{\kappa}_0$, which is the estimated short-run effect. There does not appear to be a significant short-run effect for any of the three economic variables.

The coefficient on each of the lagged economic variables is the corresponding $-\widehat{\gamma\kappa}_1$, where $\hat{\kappa}_1$ is the estimated long-run effect. We also have the estimate of γ, so we can calculate $\hat{\kappa}_1$ for each of the independent variables.

For GDP,
$$\hat{\kappa}_{1a} = -\left(\frac{\hat{\delta}_{2a}}{\hat{\gamma}}\right) = -\left(\frac{-0.54}{-0.62}\right) = -0.87.$$

For inflation,
$$\hat{\kappa}_{1b} = -\left(\frac{\hat{\delta}_{2b}}{\hat{\gamma}}\right) = -\left(\frac{-1.98}{-0.62}\right) = -3.19.$$

---

[8] The general form ECM is sometimes used with I(1) data as part of a cointegration test. See Ericsson and MacKinnon 2002.

Table 6.7 General Form Error Correction Model for Economic Popularity

D. Vote	Coefficient	Standard Error	t Statistic	P Value
L1. Vote	−0.62	0.14	−4.58	<0.001
D1. GDP	−0.53	0.44	−1.20	0.236
L1. GDP	−0.54	0.37	−1.45	0.155
D1. Inflation	0.32	1.27	0.25	0.802
L1. Inflation	−1.98	0.84	−2.34	0.024
D1. Unemployment	−5.08	5.56	−0.91	0.366
L1.Unemployment	0.84	2.33	0.36	0.719
Trend	−0.23	0.083	−2.77	0.008
Constant	35.22	18.38	1.92	0.062

NOTE: $R^2 = 0.39$, $T = 51$; T = number of time points, D1 = first difference, L1 = first lag.

For unemployment, $\quad \hat{\kappa}_{1c} = -\left(\dfrac{\hat{\delta}_{2c}}{\hat{\gamma}}\right) = -\left(\dfrac{-5.08}{-0.62}\right) = -8.19.$

We can test the hypothesis that $\kappa_1 = 0$ with an F statistic. For GDP $\hat{\kappa}_{1a}$, the test statistic is 1.96 with an $F(1, 42)$ distribution and a corresponding P value of 0.169. For GDP, we cannot reject, at the 0.05 significance level, the null hypothesis that the long-run effect is zero. For inflation $\hat{\kappa}_{1b}$, the test statistic is 7.41 with an $F(1, 42)$ distribution and a corresponding P value of 0.009. For inflation, we can reject the null hypothesis, at the 0.05 significance level, that the long-run effect is zero. An increase in inflation by 1 percentage point is estimated to produce a long-run decline in approval of 3.19 percentage points.

For unemployment $\hat{\kappa}_{1c}$, the test statistic is 0.13 with an $F(1, 42)$ distribution and a corresponding P value of 0.716. For unemployment, we cannot reject, at the 0.05 significance level, the null hypothesis that the long-run effect is zero.

As always, we want to make sure that the residuals are free from serial correlation and follow a white noise process. The Q statistic is 14.52 and is chi-squared distributed with 23 degrees of freedom. The corresponding

P value is 0.911. We cannot reject the null hypothesis, at the 0.05 signifi-cance level, that the residuals follow a white noise process. If the residuals do show signs of serial correlation, we can include lags of the (differenced) dependent variable.

We would also like to test that the residuals are homoskedastic. The Breusch-Pagan test of the null hypothesis of constant variance gives us a test statistic of 9.90. This has a chi-squared distribution with 8 degrees of freedom and a corresponding *P* value of 0.272. We cannot reject the null hypothesis, at the 0.05 significance level, that the residuals are homoske-dastic (have constant variance).

Summary

If your data are stationary, you may use the ECM, but the ADL will produce the same results. If your data are integrated and you are willing to assume that there is no long-run relationship between your variables, you can use the differenced data or ARIMA models. If you believe that there is a long-run relationship and that relationship can be represented as a cointegrating rela-tionship, the ECM is of great value. Keep in mind, though, that the single-equation ECM assumes that x_t is not a function of the cointegrating relationship—that is, x_t is weakly exogenous. If this is not an appropriate assumption, you may want to consider the multi-equation VECM (Brandt & Williams, 2007; Enders, 2004).

CONCLUSION

This text is an introduction to univariate time series analysis. In addition to covering the fundamentals of time series analysis, we have examined many of the most popular time series models: static, finite distributed lag, autoregressive distributed lag, lagged dependent variable, moving average (autocorrelated error), differenced data, GARCH (generalized autoregressive conditional heteroskedasticity), ARMA (autoregressive moving average), ARIMA (autoregressive integrated moving average), and error correction models. We have also discussed a number of approaches and guidelines to model building. These include the Box-Jenkins procedure; the general-to-specific approach to modelling; the criteria for choosing between dynamic models with lagged dependent variables and static models, with corrections for serially correlated errors; and procedures for distinguishing between stationary and integrated processes. This text has purposely avoided any single rigid procedure for model selection but has provided guidelines along the way.

In Chapter 6, we saw how the models covered in this book could be extended to the multivariate case. The fundamental concepts covered by this book provide the reader with the necessary background to pursue the study of such models. A natural extension of this text would be the material on vector autoregression and vector error correction models found in Brandt and Williams (2007). For a more detailed discussion of the material covered in this book and an extension to multivariate analysis, the reader is directed to Enders (2010). For a more detailed discussion of the material covered in this book and its relation to the analysis of cross-sectional data, the reader is directed to Wooldridge (2012). The concepts covered by this text also place the reader in a position to pursue the study of panel data analysis. Again, Wooldridge (2012) provides a good starting point, placing such analysis in the context of time series and cross-sectional analysis. For a more advanced discussion on the topic, the reader is recommended Wooldridge (2010).

I conclude with a note on software. The examples used in this text were analyzed using Stata. The same models could be run with commonly used software packages such as SPSS, R, and SAS. These packages are also capable of estimating much more complicated time series models. Some very advanced modelling techniques require much more specialized software packages such as PCGive and RATS. Bayesian time series models often require Bayesian software programs such as WinBUGS, Stan, or JAGS.

REFERENCES

Agresti, A., & Finlay, B. (2009). *Statistical methods for the social sciences* (4th ed.). Upper Saddle River, NJ: Prentice Hall.

Aguiar-Conraria, L., Bagalhães, P. C., & Soares, M. J. (2012). Cycles in politics: Wavelet analysis of political time series. *American Journal of Political Science, 56,* 500–518.

Andersen, R. (2008). *Modern methods for robust regress*ion (Sage University Papers Series on Quantitative Applications in the Social Sciences). Thousand Oaks, CA: Sage.

Andrew, B. (2010). *Media-generated shortcuts: The supply and demand of political information* (Doctoral dissertation). McGill University, Montreal, Quebec, Canada.

Angrist, J. D., & Pischke, J.-S. (2009). *Mostly harmless econometrics: An empiricist's companion.* Princeton, NJ: Princeton University Press.

Armstrong, D. A. (2008). *Measuring the democracy-repression nexus.* Paper presented to the workshop Producing Better Measures by Combining Data Cross-Temporally, University of Oxford, Oxford, England.

Arzheimer, K. (2006). "Dead men walking?" Party identification in Germany, 1977–2002. *Electoral Studies, 25,* 791–807.

Beck, N. (1989). Estimating dynamic models using Kalman filtering. *Political Analysis, 1,* 121–156.

Bentzen, J., & Smith, V. (2004). *Short-run and long-run relationships in the consumption of alcohol in the Scandinavian countries* (Working Paper No. 04–14). Aarhus, Denmark: Aarhus School of Business, Department of Economics.

Bittner, A. (2011). *Platform or personality? The role of party leaders in elections.* Oxford, England: Oxford University Press.

Bollerslev, T. (1986). Generalized autoregressive conditional heteroskedasticity. *Journal of Econometrics, 31,* 307–327.

Box, G. E. P., & Jenkins, G. (1976). *Time series analysis: Forecasting and control.* San Francisco, CA: Holden-Day.

Box, G. E. P., & Pierce, D. A. (1970). Distribution of residual autocorrelations in autoregressive integrated moving average time series models. *Journal of the American Statistical Association, 65,* 1509–1526.

Box, G. E. P., & Tiao, G. C. (1975). Intervention analysis with applications to economic and environmental problems. *Journal of the American Statistical Association, 70,* 70–79.

Box-Steffensmeier, J., & Smith, R. (1996). The dynamics of aggregate partisanship. *American Political Science Review, 90,* 567–580.

Box-Steffensmeier, J., & Smith, R. (1998). Investigating political dynamics using fractional integration methods. *American Journal of Political Science, 42,* 661–689.

Brandt, P. T., & Williams, J. (2001). A linear Poisson autoregressive model: The Poisson AR(p) model. *Political Analysis, 9,* 164–184.

Brandt, P. T., & Williams, J. T. (2007). *Multiple time series models* (Sage University Papers Series on Quantitative Applications in the Social Sciences). Thousand Oaks, CA: Sage.

Breusch, T. S. (1978). Testing for autocorrelation in dynamic linear models. *Australian Economic Papers, 17*, 334–355.

Bryman, A. E., Liao, T. F., & Lewis-Beck, M. (2004). *The SAGE encyclopedia of social science research methods* (Vol. 1). Thousand Oaks, CA: Sage.

Campbell, J. Y., & Mankiw, N. G. (1991). The response of consumption to income: A cross-country investigation. *European Economic Review, 35*, 723–767.

Campos, J., Ericssson, N. R., & Hendry, D. F. (1996). Cointegration tests in the presence of structural breaks. *Journal of Econometrics, 70*, 187–220.

Cheung, Y.-W., & Lai, K. S. (1993). A fractional cointegration analysis of the purchasing power parity. *Journal of Business and Economic Statistics, 11*, 103–112.

Commandeur, J. J. F., & Koopman, S. J. (2007). *An introduction to state-space time series analysis*. Oxford, England: Oxford University Press.

Davidson, J. (2002). A model of fractional cointegration, and tests for cointegration using the bootstrap. *Journal of Econometrics, 110*, 187–212.

Dickey, D. A., & Pantula, S. G. (2002). Determining the order of differencing in autoregressive processes. *Journal of Business & Economic Statistics, 20*, 18–24. (Twentieth anniversary commemorative issue)

Durbin, J. (1970). Testing for serial correlation in least-squares regression when some of the regressors are lagged dependent variables. *Econometrica, 38*, 410–421.

Durbin, J., & Watson, G. S. (1950). Testing for serial correlation in least-squares regression I. *Biometrika, 37*, 409–428.

Durlauf, S. N., & Blume, L. (2010). *Macroeconometrics and time series analysis*. Basingstoke, England: Palgrave Macmillan.

Durr, R. H. (1992). An essay on cointegration and error correction models. *Political Analysis, 4*, 185–228.

Enders, W. (2004). *Applied econometric time series* (2nd ed.; Wiley Series in Probability and Statistics). Hoboken, NJ: Wiley.

Enders, W. (2010). *Applied econometric times series* (3rd ed.; Wiley Series in Probability and Statistics). Hoboken, NJ: Wiley.

Enders, W., & Sandler, T. (1993). The effectiveness of antiterrorism policies: A vector-autoregression-intervention analysis. *American Political Science Review, 87*, 829–844.

Engle, R. F. (1982). Autoregressive conditional heteroscedasticity with estimates of the variance of United Kingdom inflation. *Econometrica, 50*, 987–1007.

Engle, R. F., & Granger, C. W. J. (1987). Co-integration and error-correction: Representation, estimation and testing. *Econometrica, 55*, 251–276.

Engle, R. F., Hendry, D. F., & Richard, J.-F. (1983). Exogeneity. *Econometrica, 51*(2), 277–304.

Ericsson, N. R. & MacKinnon, J. G. (2002). Distributions of error corrections tests for cointegration. *The Econometrics Journal, 5*(2), 285–318.

Erikson, R. S., & Wlezien, C. (2012). Forecasting with leading economic indicators and the polls in 2012. *Political Science & Politics, 46*, 38–39.

Evans, G., & Andersen, R. (2006). The political conditioning of economic perceptions. *Journal of Politics, 68*(1), 194–207.

Evans, G., & Pickup, M. (2010). Reversing the causal arrow: The political conditioning of economic perceptions in the 2000–2004 U.S. presidential election cycle. *Journal of Politics, 72*, 1236–1251.

Fair, R. C. (1996). The effect of economic events on votes for president: 1992 update. *Political Behavior, 18*, 119–139.

Foster, K. N., & Christian, L. M. (2008). *F*-test. In P. J. Lavrakas (Ed.), *Encyclopedia of survey research methods* (pp. 295–296). Thousand Oaks, CA: Sage.

Fox, J. (2008). *Applied regression analysis and generalized linear models* (2nd ed.). Thousand Oaks, CA: Sage.

Gill, J. (2000). *Generalized linear models: A unified approach* (Sage University Papers Series on Quantitative Applications in the Social Sciences). Thousand Oaks, CA: Sage.

Godfrey, L. G. (1978). Testing against general autoregressive and moving average error models when the regressors include lagged dependent variables. *Econometrica, 46*, 1293–1302.

Granger, C. W. J. (1969). Investigating causal relations by econometric models and cross-spectral methods. *Econometrica, 37*(3), 424–438.

Granger, C. W. J. (1980). Long memory relationships and the aggregation of dynamic models. *Journal of Econometrics, 14*, 227–238.

Granger, C. W. J., & Joyeux, R. (1980). An introduction to long-memory time series models and fractional differencing. *Journal of Time Series Analysis, 1*, 15–30.

Granger, C. W. J., & Newbold, P. (1974). Spurious regressions in econometrics. *Journal of Econometrics, 2*, 111–120.

Greene, W. (2003). *Econometric analysis* (5th ed.). Upper Saddle River, NJ: Prentice Hall.

Gronke, P., & Brehm, J. (2002). History, heterogeneity, and presidential approval: A modified ARCH approach. *Electoral Studies, 21*, 425–452.

Hamilton, J. D. (1994). *Time series analysis*. Princeton, NJ: Princeton University Press.

Harvey, A. C. (1990). *The econometric analysis of time series* (2nd ed.). Cambridge: MIT Press.

Harvey, A. C. (1993). *Time series models* (2nd ed.). Cambridge: MIT Press.

Harvey, A. C., & Durbin, J. (1986). The effects of seat belt legislation on British road casualties: A case study in structural time series modelling. *Journal of the Royal Statistical Society, Series A, 149*(3), 187–227.

Hendry, D. F. (2003). *Dynamic econometrics*. Oxford, England: Oxford University Press.

Hurwicz, L. (1950). Least-squares bias in time series. In T. C. Koopman (Ed.), *Statistical inference in dynamic economic models* (pp. 365–382). New York, NY: Wiley.

Jackman, S. (2005). Pooling the polls over an election campaign. *Australian Journal of Political Science, 40*, 499–517.

Johansen, S. (1995). *Likelihood-based inference in cointegrated vector autoregressive models*. Oxford, England: Oxford University Press.

Johansen, S. (1988). Statistical analysis of cointegration vectors. *Journal of Economic Dynamics and Control, 12*, 231–254.

Johnson, T., & Kellstedt, P. (2013). Media consumption and the dynamics of policy mood. *Political Behavior, 35*, 1–23.

Johnston, R. (2002). Prime ministerial contenders in Canada. In A. King (Ed.), *Leaders' personalities and the outcomes of democratic elections* (pp. 159–183). Oxford, England: Oxford University Press.

Keele, L. J. (2005). Macro measures and mechanics of social capital. *Political Analysis, 13*, 139–156.

Keele, L. J. (2007). Social capital and the dynamics of trust in government. *American Journal of Political Science, 51*, 241–254.

Keele, L. J. (2008). *Semiparametric regression for the social sciences*. West Sussex, England: Wiley.

Keele, L. J., & Kelly, N. J. (2006). Dynamic models for dynamic series: The ins and outs of lagged dependent variables. *Political Analysis, 14*, 186–205.

Kellstedt, P. M., McAvoy, G., & Stimson, J. A. (1996). Dynamic analysis with latent constructs. *Political Analysis, 5*, 113–150.

King, G. (1998). *Unifying political methodology: The likelihood theory of statistical inference*. Ann Arbor: Michigan University Press.

Koenker, R. (1981). A note on studentizing a test for heteroskedasticity. *Journal of Econometrics, 17*, 107–112.

Kwiatkowski, D., Phillips, P. C. B., Schmidt, P., & Shin, Y. (1992). Testing the null hypothesis of stationarity against the alternative of a unit root: How sure are we that economic time series have a unit root? *Journal of Econometrics, 54*, 159–178.

Ladner, M., & Wlezien, C. (2007). Partisan preferences, electoral prospects, and economic expectations. *Comparative Political Studies, 40*, 571–596.

Lebo, M. J., & Moore, W. H. (2003). Dynamic foreign-policy behavior. *Journal of Conflict Resolution, 47*(1), 13–32.

Li, W. K. (2004). *Dynamic checks in time series* (Monographs on Statistics and Applied Probability 102). Boca Raton, FL: Chapman & Hall/CRC.

Ljung, G. M., & Box, G. E. P. (1978). On a measure of lack of fit in time series models. *Biometrika, 65*, 297–303.

MacKinnon J. G. (2010). *Critical values for cointegration tests* (Working Paper No. 1227). Toronto, Ontario, Canada: Queen's Economics Department.

MacKuen, M. B., Erikson, R. S., & Stimson, J. A. (1989). Macropartisanship. *American Political Science Review, 83*, 1125–1142.

Maddala, G. S., & Kim, I.-M. (1998). *Unit roots, cointegration, and structural change*. Cambridge, England: Cambridge University Press.

Martin, A. D., & Quinn, K. M. (2002). Dynamic ideal point estimation via Markov chain Monte Carlo for the U.S. Supreme Court, 1953–1999. *Political Analysis, 10*, 134–153.

McAvoy, G. (1998). Measurement models for time series analysis: Estimating dynamic linear errors-in-variables models. *Political Analysis, 7*, 165–186.

Moore, W. H., & Lanoue, D. J. (2003). Domestic politics and U.S. foreign policy: A study of Cold War conflict behaviour. *Journal of Politics, 65,* 376–396.

Nason, G. P. (2008). *Wavelet methods in statistics with R.* New York, NY: Springer.

Newey, W. K., & West, K. D. (1987). A simple, positive semi-definite heteroskedasticity and autocorrelation consistent covariance matrix. *Econometrica, 55,* 703–708.

Ostrom, C. W., Jr. (1990). *Time series analysis: Regression techniques* (Sage University Papers Series on Quantitative Applications in the Social Sciences. Thousand Oaks, CA: Sage.

Ostrom, C. W., Jr., & Marra, R. F. (1986). U.S. defense spending and the Soviet estimate. *American Political Science Review, 80,* 824–839.

Pearson, K. (1905). The problem of the random walk. *Nature, 72,* 294.

Phillips, P. C. B. (1986). Understanding spurious regressions. *Journal of Econometrics, 33,* 311–340.

Phillips, P. C. B., & Perron, P. (1988). Testing for a unit root in a time series regression. *Biometrika, 75,* 335–346.

Pickup, M. (2009). Testing for fractional integration in party popularity in the presence of structural breaks. *Journal of Elections, Parties and Public Opinion, 19,* 105–116.

Pickup, M. (2010). Better know your dependent variable: A multination analysis of government support measures in economic popularity models. *British Journal of Political Science, 40,* 449–468.

Pickup, M., Andrew, B., Cutler, F., & Matthews, J. S. (2014). *The horse(race)-drawn media (band)wagon: A natural experiment on the effect of polls on campaign media coverage.* Paper presented at the 2014 Meeting of the European Political Science Association, Edinburgh, Scotland.

Pickup, M., & Evans, G. (2013). Addressing the endogeneity of economic evaluations in models of political choice. *Public Opinion Quarterly, 77*(3), 735–754.

Pickup, M., & Johnston, R. (2008). Campaign trial heats as election forecasts: Measurement error and bias in 2004 presidential campaign polls. *International Journal of Forecasting, 24,* 272–284.

Pourahmadi, M. (2001). *Foundations of time series analysis and prediction theory.* New York, NY: Wiley.

Provost, C., Gerber, B., & Pickup, M. (2009). Flying under the radar? Political control and bureaucratic resistance in the Bush EPA. In C. Provost & P. E. Teske (Eds.), *President George W. Bush's influence over bureaucracy and policy: Extraordinary times, extraordinary powers* (pp. 169–187). New York, NY: Palgrave Macmillan.

Raftery, A. E. (1995). Bayesian model selection in social research. *Sociological Methodology, 25,* 111–163.

Sims, C. (1977). Exogeneity and causal ordering in macroeconomic models. In *New methods in business cycle research: Proceedings from a conference* (pp. 23–43). Minneapolis, MN: Federal Reserve Bank of Minneapolis.

Soroka, S., & Wlezien, C. (2010). *Degrees of democracy.* Cambridge, England: Cambridge University Press.

Spanos, A. (1986). *Statistical foundations of econometric modelling.* Cambridge (UK): Cambridge University Press.

Stewart, M. C., & Clarke, H. D. (1992). The (un)importance of party leaders: Leader images and party choice in the 1987 British election. *Journal of Politics, 54,* 447–470.

Stimson, J. A. (1999). *Public opinion in America: Moods, cycles, and swings* (2nd ed.). Boulder, CO: Westview Press.

Stock, J. H., & Watson, M. W. (1993). A simple estimator of cointegrating vectors in higher order integrated systems. *Econometrica, 61,* 783–820.

Torrence, C., & Compo, G. P. (1998). A practical guide to wavelet analysis. *Bulletin of the American Meteorological Society, 79,* 61–78.

Tsay, R. S. (1988). Outliers, level shifts, and variance changes in time series. *Journal of Forecasting, 7,* 1–20.

Tufte, E. R. (2001). *The visual display of quantitative information* (2nd ed.). Cheshire, CT: Graphics Press.

Woodward, W. A., Gray, H. L., & Elliott, A. C. (2012). *Applied time series analysis.* Boca Raton, FL: CRC Press.

Wooldridge, J. M. (1991). On the application of robust, regression-based diagnostics to models of conditional means and conditional variances. *Journal of Econometrics, 47,* 5–46.

Wooldridge, J. M. (2006). *Introductory econometrics: A modern approach* (4th ed.). Mason, OH: South Western, Cengage Learning.

Wooldridge, J. M. (2010). *Econometric analysis of cross-sectional and panel data* (2nd ed.). Cambridge: MIT Press.

Wooldridge, J. M. (2012). *Introductory econometrics: A modern approach* (5th ed.). Mason, OH: South Western, Cengage Learning.

Zelle, C. (1998). A third face of dealignment? An update on party identification in Germany, 1971–94. In C. Anderson & C. Zelle. (Eds.), *Stability and a change in German elections: How electorates merge, converge, or collide* (pp. 55–69). London, England: Praeger.

Zellner, A. (1979). Statistical analysis of econometric models. *Journal of the American Statistical Association, 74,* 628–643.

INDEX

$SAGE research**methods**

The essential online tool for researchers from the world's leading methods publisher

Find exactly what you are looking for, from basic explanations to advanced discussion

More content and new features added this year!

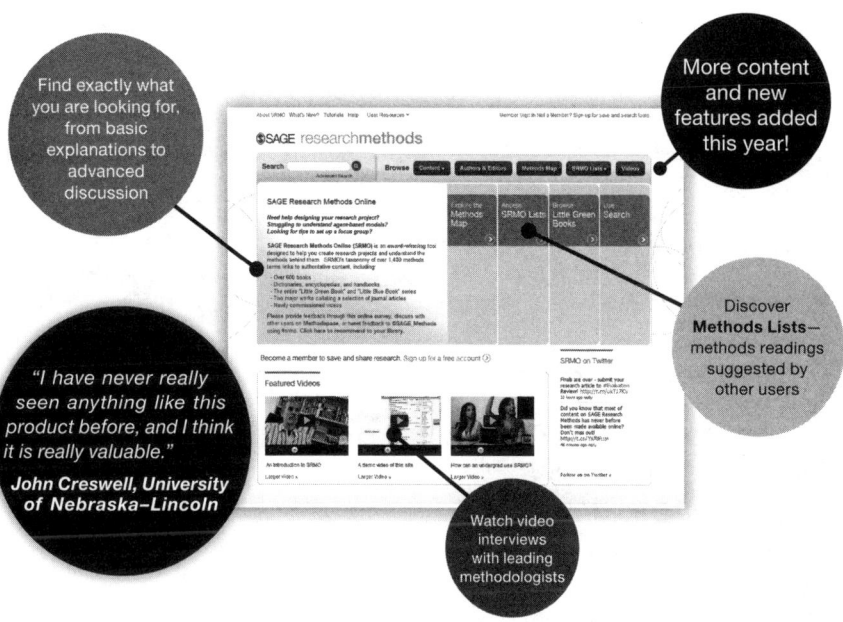

Discover **Methods Lists**—methods readings suggested by other users

"I have never really seen anything like this product before, and I think it is really valuable."

John Creswell, University of Nebraska–Lincoln

Watch video interviews with leading methodologists

Explore the **Methods Map** to discover links between methods

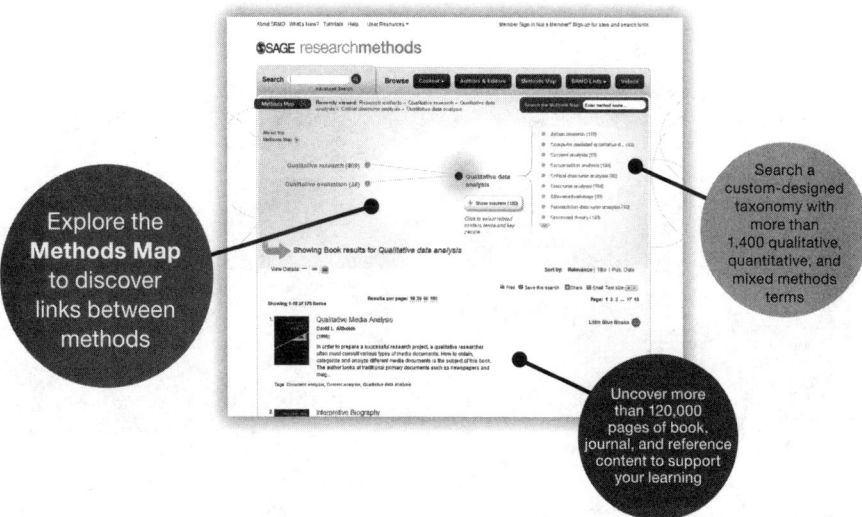

Search a custom-designed taxonomy with more than 1,400 qualitative, quantitative, and mixed methods terms

Uncover more than 120,000 pages of book, journal, and reference content to support your learning

Find out more at
www.sageresearchmethods.com